HOW TO DEAL WITH YOUR LAWYER

Richard D. Bank

HOW TO DEAL WITH YOUR LAWYER

An Inside Look

KNOWING WHEN YOU NEED ONE

HOW TO FIND THE RIGHT ONE

GETTING THE MOST FOR YOUR MONEY

BRICK HOUSE PUBLISHING CO.
New Boston, New Hampshire

For my parents, Ruth and Louis Bank,
my wife, Francine, and
my sons, Cory and Ari.

ISBN 0-931790-95-6

CONTENTS

INTRODUCTION

"Any messages, Celeste?" I anxiously asked, hoping someone interested in my services might have called.

"Na," Celeste answered. "Oh... wait a sec," she called out, as I was turning away. "There's one message, from somebody who sounded like she wanted a lawyer." She smiled, waving the pink message slip teasingly.

Snatching the paper out of her hand, I returned to my desk. Trying to regain an air of dignity, I dialed the telephone while Celeste sat with her ears perked, pretending to work. I didn't have the luxury of a door to close in my one-room office.

"Lo," a child's voice chirped.

"Hello," I responded in what I hoped was a distinguished and resonant tone. "Is your mother home?"

"Yeah... who is it?" the kid shrilled.

"This is Mr. Bank. I am an attorney."

"Yo, Mom," the voice screamed. "It's some lawya...."

"Lawya"? I was crushed. Since I was nine and watching Perry Mason bring justice to the courtroom once a week on the television screen, I dreamed of becoming an attorney. Four years of college, three more of law school, the bar exam, securing a position as an assistant district attorney, embossed announcements proclaiming the opening of my own practice, and now this... reduced to being "some lawya" to this kid. Well, surely his mother would behave with more decorum, demonstrate a degree of respect, I assured myself. Wrong.

In as coarse a language as I had ever heard, the woman recounted a scene in which a car dealer had attempted, along with an attorney brandishing a court order, to seize her car. (She had failed to keep up the lease payments.) "Parked in my own driveway!" the lady shrieked. "And dere dey were," she continued, "Bot a' dem, smokin deese beeg ceegars, dat deala and his jew lawya...." My ears singed. I heard no more. A "jew lawya...."

"Excuse me, Madame," I said, straining to control my voice, "I happen to be Jewish."

"Oh, dat's ok," she said, without the slightest trace of embarrassment, "I wanna jew lawya anyway."

During the last several decades, the image of attorneys has deteriorated to the point that I usually grin sheepishly when conceding my membership. I have grown accustomed to "lawyer jokes" aimed at me across a dinner table or at a cocktail party.

"So you're Henry's lawyer?" the bald banker in black asks. I nod in acknowledgement. "Hey, wanna hear a good one?" The chubby financier is already chuckling. "Sure," I say. What else can I say? The banker takes a belt from his highball. "Why is it that only the lawyer survived when a fishing boat went down in shark-infested waters?" "I don't know, how come?" I ask. "Professional courtesy!" The banker guffaws. I snicker politely. Heard it before.

The problem is not the personal distress attorneys may suffer at social gatherings. The real dilemma confronts the public at large.

One solution which many appear ready to adopt is to follow the advice of one of Shakespeare's characters: "First thing let's do—let's kill all the lawyers!" It is almost as if a neon sign hangs over the head of every attorney in America flashing a warning which reads: "Avoid if at all possible, or use sparingly."

There is a saying. You must have heard it. "He who is always his own counsellor will often have a fool for his client." It is true; believe me, I know. This is why I always retained outside counsel when I was president of a mortgage company. Thus having sat on both sides of the table, as attorney and client, I experienced the worst of both worlds.

Left to fend for themselves, most people would not choose self representation. However, there is a wide array of books available on "How To" avoid using an attorney. In addition, legal software packages have been introduced into the marketplace which enable anyone with a compatible personal computer and printer to prepare all sorts of legal documents. Some of these books and computer programs are better than others, but they all pose a potential danger to the unsuspecting layperson.

Why? Laws are constantly changing. Ever see the law library in an attorney's office? I am often tempted to tell my clients I have committed it all to memory, but that would be a lie. In addition to the new books comprising recent court decisions, there is generally at the back of each text, even the musty old ones, an insert with a current update to that volume. These pocket parts are sent at regular intervals to law firms by publishers or distributors.

The law undergoes transformation in two ways. First, the courts alter the law as they clarify or even overturn previous decisions, and make new law when a fresh set of circumstances presents a novel issue. Second, the legislative bodies enact laws, repeal old ones, or modify others. Thus, it means nothing to learn the law unless one keeps abreast with the changes.

While some "How To ... Without A Lawyer" books may prove useful in specific situations, they can also be either misleading or totally unnecessary. For instance, no one should write his or her own will. Most attorneys charge nominal fees for the preparation of simple wills, and complex wills involving estate planning require an attorney. The fee for standard husband-and-wife wills is about a hundred dollars and sometimes even less. So why write your own will and take a chance on making a mistake, when you save little money, if any (you did have to buy the "How To" book or computer program)?

Another fallacy with "How To ... Without A Lawyer" books lies in their attempt to instruct the reader in the law. Learning the law, which is more than mere memorization, is no easy matter, and only certain types of people are cut out for it. Possibly embedded within you is a keen legal mind yearning to be extricated. Then again, maybe not.

What lawyers do is apply the law to a set of facts. Obviously, in your given instance, you understand the facts better than anyone else, so you may think this is a straightforward task. But what you don't know are the myriad things that can go wrong and what courses of action should be considered to avoid them. From experience, and not necessarily wisdom, lawyers usually do.

For example, the average person engages in two or three real estate transactions in a lifetime. I have been involved in thousands, and yet I still see a new twist here and there, which if left unchecked can cause tumult and trouble. However, because of my background, I can at least anticipate what can happen and therefore plan to prevent it. All the "How To" books can never give you that.

In addition to avoiding lawyers altogether, the other mistake people make is to delay contacting an attorney until it is no longer worthwhile. Would you ask your accountant to review your tax returns after they were filed and you just received an audit notice from the IRS? Would you solicit a second opinion from a doctor after the operation was already performed? Probably not. Yet, thousands of calls just like that are made every day to lawyers from people requesting representation.

"I just bought a house.... Certainly, I signed the agreement.... When? Why, months ago.... Yes, I did send back my mortgage commitment.... Of course, it was signed.... Settlement is scheduled for next Monday.... Do I need a lawyer to go with me?"

That's a call I get at least once a month. "Sure, I'll be happy to go with you to the closing, hold your hand and smile knowingly, if it will make you feel better. Now, there remains the matter of my fee ...." In effect, this is the answer most attorneys will give. I don't. Not that I'm a great humanitarian, it's simply that I can't justify charging a fee large enough to make it worth my while to attend a one-hour settlement.

What I do provide is a free lecture on why I should have been called before the agreement was executed. I emphasize that buying a house is generally the largest and most important financial transaction the average person makes. Yet, in the vast majority of cases, it is done without a lawyer, or an attorney is brought in at the end when very little can be accomplished.

Given what I have said thus far, there are two conclusions to be drawn:

Conclusion One. When an attorney is required (Chapter Two) self-representation and a "How To" book is no substitute for the *right* lawyer (Chapter Three).

Conclusion Two: Don't wait until it is too late. The time to discuss a matter with an attorney is at the beginning or sometimes even earlier, when you are just thinking about doing whatever it is you may want to do.

This book is intended to enable you effectively to employ an attorney, when you need one, as efficiently and inexpensively as possible. That's the whole of it. If you don't want to know this and are still insistent upon going it on your own, stop right here and exchange this book for "How to Represent Yourself in Bankruptcy," which I suspect may come in handy by the time you are done. On the other hand, I would be delighted to have you join me through the stages of the attorney-client relationship in the pages to follow.

CHAPTER ONE

Lawyers: They're People Too... Or at Least They Were Born That Way!

I once provided a student intern with Kafka's *The Trial* as an introduction to our legal system. In that novel, K, the principal character, was forever frustrated trying to work his way through the muddle of the law court in Kafka's book. Not informed of the charges nor the "evidence" against him, K was helpless to defend himself. At the end, awaiting the executioner's blade, K beseeched the surrealistic tribunal which imposed the sentence: Where is the Judge whom I have never seen? Where is the High Court which denied me?

The mystique which shrouds jurisprudence is the chief weapon in an attorney's arsenal, and like Kafka's protagonist, the public is refused access by those who jealously guard the courts of law. This helps lawyers promote the idea that they are better than everyone else because they understand the law and you don't. That's nonsense.

My plumber can install a hot water boiler. I can't. The auto mechanic can get my car to run while I peer over his shoulder. When my body is invaded by invisible bacterium all I can do is drag my ailing self to the doctor's waiting room while he scribbles indecipherably on a pad to prescribe just the right pills to make me well. Now my plumber, auto mechanic, and doctor (perhaps we should exclude the doctor; physicians are a lot like lawyers) don't go around pretending they're better than me. So why should you accept that attitude from a lawyer?

The thing to bear in mind is this. No matter how overbearing lawyers may be, the answer is not to avoid them. Instead, make them work to your advantage and don't allow yourself to be placed in a subservient position in the attorney-client relationship.

1

In order to accomplish this, you must understand what makes lawyers tick. So, let's take a look and see what kind of metamorphosis turns an ordinary, everyday human being into a lawyer.

All lawyers must graduate from an accredited law school. Unlike virtually every other profession, however, there is no required undergraduate curriculum to enter.Therefore, your lawyer's legal training began and ended with three years in law school. While this may be a demanding time, it is only three years, less summers, vacations, and the like. Yet, people often assume lawyers are a source of sage advice on practically every subject. I am frequently asked for financial guidance. Or I am asked, "Should I buy this house? Should I sell my business? Should I leave my wife?" In some cases, I may offer a qualified opinion on real estate or commercial matters but I do this because of my business experience, not because I am an attorney.

The training of attorneys has little which would qualify them to impart words of wisdom on anything other than the law. So, unless you have good reason to believe your lawyer possesses the expertise and knowledge in a given area, don't ask for it.

After graduation from law school and passing the bar exam, lawyers can practice law in a number of ways. However, our only concern here is the lawyer who represents the private client - you. Understanding how this is done is of critical importance to you in handling your lawyer.

Surgeons operate in hospitals. Entertainers perform on the stage. Lawyers set up shop in law firms. And law firms operate just like any other business, driven by the profit motive. Every attorney bears the same burden, the need to earn dollars. Whatever the circumstances, the pressure to make money is forever present. Unless your attorney is independently wealthy, this pressure will affect the relationship you have with your lawyer. Let's see how it works.

You have engaged an attorney who specializes in personal injury matters. The case was taken on a contingent fee basis (to be discussed in Chapter Four). However, yours is not a substantial matter. Ms. Prudence Juris, your attorney, has several six figure cases awaiting trial. She doesn't want to jeopardize her position on those matters by initiating negotiating overtures just yet. Hence a cash flow problem.

As luck would have it (not necessarily for you), the insurance adjuster offers $7,500 to settle. How will Ms. Juris convey the settlement offer to you? Will she recommend it? Will she tell you what she thinks the case can be settled for a year from now?

For another example, consider this. You are divulging to a junior partner in a prestigious firm the circumstances involving a ten thousand dollar investment which went sour. He advises you there is a basis for a law suit against the promoters of the scheme. What he doesn't disclose is that in five months he comes up for review to make senior partner, and he is significantly behind on the number of new clients he has brought to the firm. He really needs you and your case.   Does he tell you the legal fees may exceed the amount you lost in the bad investment? Does he suggest an investigation be made to determine whether the defendant has any assets should you prove successful? Or does he just recite the law and say your chances are better than good you will win?

The crash of a trash can outside your bedroom window rouses you from a much needed sleep. For the fourth night in a row your next door neighbor, the Whipples' horny tomcat is on the prowl. That's it! Now you had it! Add that to the branches from the Whipples' dying maple hanging on your side of the property line and the water discharged from their downspout onto your lawn causing the basement to flood .... The only thing which gets you to sleep is the satisfaction in knowing you can telephone your lawyer in the morning and then sue the pants off old man Whipple!

It just happens that at the very instant your call is put through, your attorney is scanning the ads for summer homes at the shore. He listens politely while you storm on about the Whipples, their dying maple, sprouting downspout, and horny tomcat. Neighbor disputes can get nasty, and when accompanied by litigation, they can also be very expensive. The end result is rarely satisfactory to either party; therefore, staying out of court is an important objective. It is a time when attorneys must interject a degree of wisdom and constraint.

Your lawyer knows this very well. He also knows it is not what you want to hear. "I don't care how much it costs me!" You scream. While you rage, your lawyer ponders a nice little house just a block from the beach.

These illustrations demonstrate the subtle effect financial considerations have on lawyers, who are subject to the same human frailties as everyone else. Take the recent case of JW.

JW was a prominent attorney who became involved, both financially and romantically, with the madam of a house of prostitution disguised as a massage parlor. Whether to protect his investment, or acting out of passion, as JW claimed, he bribed a police inspector to prevent the establishment from being closed. During an FBI investigation into police corruption, this mischievous escapade of JW's was unearthed and disciplinary proceedings commenced.

To the surprise and outrage of many, the State Supreme Court's decision was not disbarment but to suspend him for four years, on the following rationale: "In this case, there are human beings who are fallible, who become embroiled in relationships and do foolish things."

There you have it. Inscribed in the annals of American Jurisprudence it has been decreed that yes, lawyers are human.

Of what concern is this to you, the client? For one thing, your lawyer's human vulnerability can affect the quality of legal representation you receive. A 1990 survey by the American Bar Association has disclosed that thirteen percent of those lawyers questioned admitted to having six or more alcoholic drinks a day. This compared to less than one-half of one percent who said they drank that much in 1984 and about one percent of the general population who concedes to five drinks or more a day.

In other words, your lawyer may be in a stupor when rehearsing the closing arguments for your trial or may be drunk while dictating your employment contract. Keep a discerning eye on your attorney's appearance and general demeanor.

The second and more important reason for you to look upon lawyers as being mortal is this: So that you won't be afraid of them.

If you acknowledge that attorneys really are only human after all, then you will never be afraid of one. That's half the battle, not being afraid of lawyers. It's a lot like the saying: "The only thing we have to fear is fear itself " (said by Franklin D. Roosevelt, another attorney who became president).

CHAPTER TWO

When Do You Need an Attorney?

The sun sparkled on the clear sheen of the Caribbean. Stretching my legs, I felt the surf gingerly lap at my twitching toes. In one hand, I held a drink from which I occasionally sipped through a red and white plastic straw. With the other hand, I waved languidly at a flock of voluptuous women strolling by on the beach.

Abruptly, a phone rang. My head bounced off the pillow. I tried to focus my bleary eyes on the telephone. With unsteady hands, I bobbled the receiver, clunking it on the headboard before getting a grip on it. My heart pounded in anticipation of the dreadful news I was certain was forthcoming. After all, I thought, as I gazed at the glowing digital clock, who would call me at four in the morning... four in the morning!

"Hello," I rasped.

"Reecherd." The voice at the other end sounded vaguely familiar. "Eet's me, Tony." Tony? My brain was still clinging to the Caribbean. A stammering halfhearted effort to apologize for telephoning at this hour provided the time I needed to determine that it was Tony the painter, a client who in exchange for occasional legal services, gave me bargain prices for house painting.

"What is it, Tony?" I grumbled.

As it turned out, Tony also had been awakened during the early morning hours. But instead of a phone, he heard the rattle of chains being attached to his truck by the county sheriff, who was in the process of towing the vehicle away.

"Reecherd, you can't letta him do dis ... I'a need my truck to paint!"

"Calm down, Tony," I said. "What is this all about? A sheriff doesn't just appear from nowhere and attach property." In the next few minutes, I learned from a rambling, indignant discourse that many months ago, after ignoring repeated letters demanding payment on outstanding bills due a paint supplier, Tony was sued. Naturally, the Complaint served by the constable was likewise ignored as well as the subsequent notice that judgement had been entered. Being consistent, Tony also tossed out the Writ of Execution levied upon his car and truck. Yet, despite all this, Tony was convinced he was not to blame for this outrage, but instead, that "no a'good lady lawyer" who represented the paint supplier.

Somehow, I persuaded the sheriff to execute on the car but leave the paint truck so Tony could continue to work.

The next day, I reached an agreement with the attorney upon whom Tony had hurled so many derogatory epithets. Tony's car was returned within the week, and the indebtedness was paid in monthly installments.

Although Tony and I remained on good terms, he no longer required a lawyer, nor I, a painter. One day, several years later, I received a phone call from an attorney claiming to represent Tony who requested that I provide information from one of his old files. It is not unusual for clients to change attorneys (Chapter Five). What shocked me was that Tony's new lawyer was the same woman who arranged for his truck and car to be hauled away in the middle of the night! When I called to confirm this with Tony, he explained he wanted an attorney who would be tough.

And so the story ends. Tony got his car back. The painting supplier got paid. The tenacious attorney earned a nice fee and later got Tony as a client. And I ... well, I returned to the Caribbean the very next night.

It was obvious, even to Tony, that he needed a lawyer. The problem for Tony, and especially for me, was that this did not occur to him until four o'clock in the morning when there was not much an attorney could do. Hence, the importance of the question, "When do you need an attorney?"

Remember the recurring telephone calls I received regarding representation at real estate settlements? Other than the timing, a little late like Tony, what else was wrong? Just from whom were these people seeking an answer?

Would you ask an automobile dealer to tell you if you need a new car? Or an insurance agent if you have enough life insurance? Attorneys have a business to run (Remember Chapter One). So, asking an attorney if you need a lawyer isn't always the best way to find out.

There are exceptions. If you know or previously have been represented by the attorney, or he or she is very successful and does not need unnecessary cases to earn a fee, you may get good advice. Otherwise, you should rely on your own common sense and what you learn in this chapter.

Litigation: Something You Go into as a Pig and Come Out as a Sausage

I recite this definition to clients whenever litigation looms on the horizon. It is something to be avoided, and all reputable attorneys will attest to this. In a word, whenever possible, compromise, no matter how distasteful it may be. Participating in the combative arena of the courtroom is like shooting dice on the craps table, except the stakes can go beyond your initial wager. Litigation is costly, time-consuming, confusing, and exasperating. More often than not, the only winners are the lawyers. However, if after due deliberation you decide to take legal action against someone, or find yourself at the other end of a lawsuit, you will almost always need an attorney.

The exception to this rule generally involves small cases within the domain of the local justice of the peace, municipal court or small claims courts usually confined to landlord/tenant disputes, minor criminal matters, including preliminary hearings and motor vehicle violations, and civil controversies over amounts not beyond certain limits. These courts are specifically designed for the layperson and conducted accordingly. In fact, in many areas of the country, the magistrates are not lawyers.

Even so, there are occasions in such courts when you would be better served by an attorney. For instance, if a question turns on a legal issue rather than the facts, an attorney's knowledge of the law can help your case.

No matter how trivial it may appear, every criminal matter must be taken seriously. The ramifications of a conviction or guilty plea may be more than you think. At the very least, consult an attorney before your hearing.

Often these courts are consumer-oriented, and if your business becomes involved in a suit at this level, it is wise to have an attorney by your side to remind the district justice of the law. In fact, you may not even have a choice, since in some jurisdictions corporations are required to be represented by counsel.

Finally, having an attorney can work to your benefit in more subtle ways. The judge and your lawyer may be on friendly terms, especially in rural regions where there are not many members of the local bar. It is possible the magistrate is impressed you took the matter seriously enough to retain an attorney. Certainly the fact the case has already cost you financially will not be lost on the judge who may feel the money paid to the lawyer is punishment enough!

The important thing to bear in mind is that there is a right of appeal from such courts. So if you lose, you can hire a lawyer and go to "a higher court." However, if in doubt, retain counsel.

Putting aside these district, municipal, and small claims courts, the rule is hard and fast. When in court, you need a lawyer. The corollary is - do not procrastinate. Never, never, never, ignore a lawsuit. The consequence of failing to answer a Complaint is disaster.

Business, Corporations, Partnerships: The Great American Dream

Statistics have shown that ninety percent of all new businesses eventually fail. Yet this does nothing to dissuade people from plunging into the world of commerce with the goal of making millions and being their own boss. Not far from this credo is the adage, "An ounce of prevention is worth a pound of cure." This sets the standard for employing lawyers in connection with your business. If utilized properly, legal advice as you chart your affairs and operate your enterprise can avoid numerous problems, such as litigation. Here is an example of how this works.

Unlike the solely owned proprietorship, partnerships and usually corporations consist of two or more owners. In the beginning, the last thing the partners or shareholders consider is that all might not work out. Yet then, while peace still prevails, is the time to plan for it, with a Stock Restrictive Agreement, also called a Buy-Sell Agreement, or clauses in the Partnership Agreement covering this contingency.

"Oh, but that's not necessary. We'll never need that!" Clients often say when I recommend it at the initial meeting. That is when I recount the experience of Manny and Max, who were not only partners in a thriving business but brothers as well. Between Manny and Max, things deteriorated to the point where, although in the same office, neither spoke to the other and they communicated through their employees.

Fortunately, closing the company and liquidating the assets at much less than value was avoided because of an all night negotiating session in my office. One brother sat in the library with the company's accountant. The other brother paced the halls, his wife goading him on. Both insisted that I, counsel to the corporation, be the only attorney present. And so I shuttled back and forth between the two until an agreement was hammered out. One brother bought the interest of the other at a fair price and both were spared litigation.

This tale is usually convincing proof to all my clients that such agreements are necessary. But do not take this to mean you should ask your attorney how to conduct your business. (Remember Chapter One?) Nevertheless, there are occasions when the services of your lawyer are almost always appropriate. To highlight some of them, I have prepared the checklist shown on the next page.

Consider an attorney to be a worthwhile cost of operations. Make it a line item in your operating budget. Naturally, you will want to do this as efficiently and economically as possible, which can be accomplished by selecting the right attorney (Chapter Three) and establishing the ground rules (Chapter Four).

> **When You Need a Lawyer**
> - Formation and type of business.
> - Regulations and licensing requirements.
> - Preparation, negotiation and execution of contracts, franchise agreements, leases, letters of intent, etc.
> - Buying or selling a business or any asset not in the ordinary course of your trade.
> - Patents, trade secrets, copyrights, and trademarks.
> - Extending warranties to the public.
> - Extending credit to the public which triggers compliance with Regulation Z and Truth-in-Lending.
> - Employee relations, including hiring, discipline, evaluations, promotions and termination.
> - Bankruptcy proceedings—especially reaching the decision whether to file or work out an arrangement with creditors.

Wanted, Dead Or Alive : Bounty Hunters In Three Piece Suits

"Hell knows no fury like a lawyer on a contingency fee." So said U.S. Senator John Heinz when introducing a bill allowing attorneys to institute civil suits against savings and loan officials. He went on, "I can't think of anybody who deserves more to be hunted down by one of these lawyers than the thieves who have looted the federal Treasury."

If you are the plaintiff in a personal injury claim, you need a lawyer. This does not mean every time you suffer a misfortune or accident you should run to the lawyer with the largest ad in the yellow pages to see if you should sue. A legal opinion is an ingredient in making this decision, but your good sense and conscience must also be brought into play. To put it bluntly, if you were not hurt don't let a lawyer talk you into suing.

However, where you have been injured as the result of someone's negligence or action and you decide to prosecute the matter, you should have a lawyer.

Domestic Relations: Until
Death (or Lawyers) Do Us Part...

Understandably, people are not often disposed to seek an attorney in divorce, the most personal aspect of life. Another reason is the perception that currently amiable, or at least civil, behavior will deteriorate to outright hostility if attorneys are added. Nevertheless, lawyers are essential in matters of separation, divorce, and custody of children.

On the other end of the spectrum are those who plan ahead. This involves an agreement spelling out who gets what if marriage partners split. When signed before the smitten couple exchange vows, this is called a Pre-Nuptial Agreement; if after, a Post-Nuptial Agreement; in either case, they need a lawyer. In fact, these agreements are generally unenforceable should one party not be represented by legal counsel.

Perhaps here, more than in any other area of the law, selecting the right attorney is crucial. In a contentious divorce, an aggressive attorney is appropriate, but otherwise a more sensitive, wise, and conciliatory one is far better. (This is explored further in Chapter Three.) In any event, the rule is clear: Unless there are no children; unless the apportionment of marital assets (which are minimal) is not in dispute; and unless both parties are in total accord on all issues, including the desirability of the divorce; a lawyer is required.

This is so for many reasons. For one, emotions and irrational thinking need to be tempered and tamed, a service attorneys can furnish especially in the matter of child custody, where too often children become the pawns of their parents' passions. Attorneys can explain legal rights and obligations to husband and wife in order to conduct fair negotiations and achieve an equitable settlement. This is often necessary to prevent one spouse from literally "giving the house away," which happens where one party functions as a doormat to the other. Attorneys are also useful in pointing out tax ramifications, which may otherwise go unnoticed. And ultimately, should a hearing or trial be necessary, the primary rule, "When in court you need a lawyer," applies.

Criminal Cases: You Have A Right
To An Attorney—But, Do You Need One?

In deciding *Gideon v. Wainwright*, the United States Supreme Court determined that a person accused of committing a crime cannot receive a fair trial without an attorney.

Therefore, if the defendant is indigent, it is the responsibility of the State to furnish counsel. The Court reasoned that given the resources at the State's disposal, including the police and the prosecutor's office, it would be grossly unfair to expect a poor and often semi-literate individual to stand alone under the accusatory arm of the Commonwealth. Hence, to the rescue, comes the underpaid, overworked, frequently dedicated though sometimes cynical public defender.

But what choice do you have when confronted with a criminal prosecution if you are not wealthy enough to pay for a prominent defense attorney, yet you have sufficient assets to rule out a public defender or court-appointed lawyer? Do you hire the best attorney you can afford, or go it alone? The answer is simple.

Heed the Supreme Court, no matter what you may think of it. No one, absolutely no one, is on your side. Not the arresting officer who tries to make light of your dilemma, nor the patronizing prosecutor indulgently explaining the charges, nor the seemingly sympathetic judge peering down from the bench reminding you of a stern but compassionate and caring parent who would never do you ill. Under this pretense of protection and concern, you sit solitary and abandoned at the seat of justice unless you are flanked by a lawyer.

Anyway, you know the rule by now. When in court, you need a lawyer. This goes double for criminal cases. Not only do you need a lawyer but you need the best you can afford.

Real Estate: The Landed Gentry—Everyone's
Dream...Or Nightmare

In Woody Allen's movie, "Love and Death," the father, an elderly and frail Russian peasant type, passes on the totality of his wisdom to the urchin Allen with one word: Ground. To demonstrate this point, he carries a tiny plot of earth in his pocket, representing his belief in the essential value of property.

The importance attached to land over the centuries has never diminished, and holds a paramount place in our society. Everyone, at least sometime, will feel the effect of real property law. This can occur in a number of ways: as a tenant or landlord; as an investor; as a buyer or seller of a house; in an argument with a neighbor over where a boundary fence properly lies; when your elderly parent gifts the family home to you but retains the right to live there; and so on.

While you don't need a lawyer every time you become involved with real estate, there are occasions when you do. Remember the phone call I received from the person requesting representation just days before the closing? What might have gone wrong earlier which could no longer be corrected?

For one thing, there was an Agreement, contingent upon the buyers obtaining a mortgage commitment. This, the buyers thought, meant that if a mortgage could not be obtained, the deposit money would be returned. Wrong. A mortgage commitment was issued which satisfied the requirement. However, there was a condition in the commitment that the buyers sell their present house, which they were unable to do by the time of settlement. No mortgage, no settlement, and no return of deposit.

Another scenario. Ned and Nancy are gazing into each other's glowing eyes across the conference table, where they are closing on their newly constructed "dream" home.

Finances were tight going in but they made it, almost to the penny. It is December and the builder explains the driveway cannot be blacktopped until spring nor can the lot be graded and seeded. "That's ok," say Ned and Nancy. They like the builder and trust him to finish the job.

"Wait just a minute," interrupts the settlement clerk, clearing her throat. "The mortgage lender's instructions require that everything be completed." The real estate agent offers to call the mortgage company. Anxiously, Ned and Nancy eavesdrop on the telephone conversation between the agent and the mortgage loan officer. Hanging up, the realtor exhibits a matronly beam. "No problem, the bank will accept a four thousand dollar escrow."

Ned and Nancy look at the builder. He, too, is smiling as he points to the Agreement which states in paragraph twenty-three that any escrow required by the mortgage lender is the responsibility of the buyer.

The builder is smiling. The real estate agent is smiling. The settlement clerk is smiling. Even the mortgage loan officer at the other end of the phone line is probably smiling. It must be contagious! Everyone is smiling but Ned and Nancy. They don't have the four thousand dollars.

In a real estate transaction, much like the defendant in a criminal prosecution, the buyers have no one to look out for their interests unless they have an attorney. Therefore, the following directive: When purchasing real property, unless you are in the business and trained accordingly, you need a lawyer.

Sellers, on the other hand, have more flexibility. Real estate brokers are *their* agents and must represent them, though this is no substitute for a lawyer. In addition, such transactions are less complex for sellers, who normally go to the closing, sign the deed, and receive the proceeds. This is not to suggest that retaining an attorney is ill-advised, only that it is not mandatory.

However, one thing applies to buyer and seller alike. The time to hire a lawyer is before signing the Agreement. In fact, this extends to every contract you execute. Far better to have your attorney explain the lease prior to discovering that the small print tells you your adoring cocker spaniel, Taffy, cannot live with you in your new apartment, or that zoning laws prohibit you from raising bullfrogs in the back yard.

Wills and Estates: Death and Taxes, Life's Certainties—An Accountant for One, A Lawyer for the Other

I'm twenty-five, married, just passed the bar exam, a bright future lies on the horizon and I decide it's time to write my will. That's what law school does to a person. Trained to anticipate the pitfalls and problems clients can encounter puts a chill on things. Yet, on the other hand, this has made me perceptive, objective and always rational—also unable to fathom my wife's tears when I expound upon our wills.

People tend to avoid the unpleasant and, with certain exceptions, contemplating one's own death is unpleasant. Nevertheless, almost everyone should have a will. As I explained in the
Introduction, it is foolish to do it yourself when attorneys charge
such reasonable fees for this service.

If you die without a will, you die intestate. This has nothing
to do with how you are buried. It means your assets will be
distributed in a manner prescribed by the state, which can result
in a share passing to your older sister, whom you have hated ever
since you were little. It also means that a judge will determine
who shall care for your children if they are minors. Should these
or similar issues be important considerations, have a will prepared.

While I do not advocate spending every waking hour planning
for your death, or freezing all your assets in irrevocable trusts
which, though saving estate taxes, may leave you cash poor,
everyone should be aware of estate planning. An attorney can
ascertain if this is even a concern, and then suggest what should
be done about it.

Therefore, the rule is absolute. Contact an attorney and
determine whether you need a will. If so, have one drafted. What
happens after you or a loved one dies is another matter.

Before the assets of an estate can be dispensed, the will must
first be probated, where its authenticity can be assessed. Here
your executor should hesitate before retaining counsel. Attorneys' fees can be considerable and often are based upon a sliding
percentage of the estate's size so they are relatively substantial
for smaller estates.

Because legal fees can be significant, your executor may want
to consider probating the will without an attorney. Often the
personnel in the Register of Wills Office or similar division in the
county court house are helpful. If necessary, an accountant can
prepare the tax returns. Assuming typical assets, many of which
are personal belongings, disposition should not be complicated.

Should you be named executor of an estate, visit two or three
attorneys. You are under no obligation to choose the lawyer who
wrote the will. How to pick the right lawyer and set the fee will be

discussed in subsequent chapters. If the estate is sizeable and the attorney's fee reasonable, retain counsel. Otherwise, do not be reluctant to probate the will yourself, so long as you have the time and inclination.

Taking On The IRS—When Do You Need A Tax Attorney?

"Hey Guido... You'a get mee da best, and I'a mean da best damn Taxa Lawya dere is! You'a gotta dat, Guido?"

Unfortunately for Mr. Alfonse Capone, there wasn't an attorney in the world (though Guido died trying to find one) who could have saved Scarface from the clutches of the federal prosecutor's office and a guilty verdict on tax evasion charges. While you certainly should not emulate Al Capone's haphazard manner of reporting (or not reporting) income, hiring a tax attorney if you find yourself heading for Tax Court is a good move. Do it.

But what about other tax situations? When do you need a tax attorney and when will an accountant suffice? The fact is most people never need a tax attorney, and unless you become involved in extremely sophisticated financial transactions or your annual tax bill to the government moves into the six figures, there is no reason to have one.

I am frequently asked by my small-business clients whether they need a tax attorney. I answer by repeating what another of my clients, the owner of a small chain of restaurants, once said: "To succeed in business, you need three things: First the head for it, and after that, a good lawyer and accountant."

There you have it—or hopefully you do. Your lawyer and accountant will be able to tell you when to spring for the three hundred dollar an hour tax attorney. Generally, your accountant will be able to handle all of your tax concerns. When more is required, rely on this team of professionals to advise you accordingly. Hiring a tax attorney when it isn't necessary is like buying a Ferrari to drive up and down your driveway. Other than impressing the neighbors, what's the point?

Everyday Life: Contracts and Consumer Affairs—When Do You Need A Lawyer?

When president of a mortgage company, I had dealings with bankers in the rural regions of Pennsylvania. I soon learned that they were people of their word, and although contracts were necessary, a handshake between us was always sufficient guarantee. Wherever such customs still exist be grateful and don't complicate it with a lawyer.

However, once agreements and understandings are reduced to writing, it's another story. Should the document be simple and concise and you have no question concerning the language, or you are familiar with what is a standard instrument in a particular trade, you do not need a lawyer. Otherwise, you do.

When the backslapping salesman dismisses the "routine/just a matter of form" contract with a sheepish grin, take the time to study it. If you don't understand it, have an attorney read it, especially when there is an inconsistency between what you are told and what the agreement appears to say. This applies to buying a car or arranging for an in-ground pool to be dug in your rear yard or purchasing a "lifetime" membership at a health club.

Do not be taken in by the nonchalance of someone casually sliding a document with parts in infinitesimal print under your nose, who offhandedly mentions it is not subject to change so why bother to read it. Beware the assurance that a particular paragraph is never enforced "anyway" or it was meant to be struck and will be, after you sign. Pay no attention to the "don't you trust me?" expression on the merchant's face.

Carefully reviewing a legally binding instrument is not an affront to anyone's integrity, even with that person heavily breathing over your shoulder and brandishing a pen in your line of vision. And, though well-intentioned, it is possible that the salesperson doesn't understand the contract, let alone be in a position to explain it.

Given the choppy waters of "everyday life," there is no fixed heading to signal when it is appropriate to retain an attorney. Nevertheless, I suggest two bearings upon which you may chart your course. First, the more consequential the transaction, the greater the likelihood you will need a lawyer. Such instances generally arise where a large sum of money is involved which you pay immediately or for which you become contractually obligated. Naturally, this is relative, but as a rule of thumb, if the money at stake is more than you can afford to lose, you have reached this level.

Second, if something is not "sitting right" in your gut or you have "second thoughts" about proceeding, rely on your instinct. See a lawyer. Yet, on the other hand, do not execute an important contract without legal advice merely because the other party has "a friendly face" and you feel good about it.

Like telephones, the law reaches out and touches everyone in many ways and at many times, though not always innocuously. Often, resembling a virus invading your body, you won't even realize it is happening. But there are clues, some of which I have discussed here. Wherever possible, take preventive action and consult an attorney. Should this fail to occur, at the first sign of a symptom, such as a lawsuit, retain counsel just as you would see a doctor to cure your illness.

Now that you know when you need a lawyer, the question becomes how to find the right one. So, on to the next chapter, "Selecting An Attorney."

CHAPTER THREE

Selecting an Attorney

Picture this: Two men attired in Brooks Brothers business suits are putting the finishing touches on their meal with a bottle of brandy and cigars. Leaning back and blowing out smoke, one remarks pompously to the other, "If there's one thing I've learned in all my years in the business world, it's this: Never see a lawyer without seeing another lawyer first!"

The problem with this adage is that it begs the question: do you first have to see a lawyer about the lawyer you intend to see concerning the lawyer you intend to hire, and so on? Hence, Chapter Three, your guide to the task of selecting an attorney.

If you were having trouble with your sinuses, would you go to an orthopedist? That would be foolish. Similarly, you don't hire a personal injury lawyer to incorporate your business, nor should you retain a corporate attorney to defend you in a criminal matter. This may seem straightforward enough, but it's amazing how people often follow just this approach in choosing an attorney. For an extreme example, consider the case of Julius and Ethel Rosenberg.

In 1950, Mr. and Mrs. Rosenberg were accused by the United States with violating The Espionage Act of 1917 for allegedly transmitting the secret of the atom bomb to the Soviet Union. If this does not sound serious enough, take note that their crime was punishable by death.

Mr. and Mrs. Rosenberg were very intelligent individuals, so you would expect that they would retain the best criminal defense attorneys available. Wrong. Julius was represented by Alexander Bloch, a well respected, seventy-four year old business lawyer, who specialized in the sale of bakeries. Ethel hired Mr. Bloch's son, Emanuel, who had handled several civil rights cases, but whose experience in criminal law was almost nil.

So stood Julius and Ethel Rosenberg against all the might the United States government could muster in, perhaps, the most famous spy trial of the century. Defending them were two lawyers —dedicated, compassionate, and self-sacrificing (so much so that the case probably killed Emanuel)—but who, together, knew next to nothing about criminal law. It may not have affected the outcome (they were electrocuted) but the Rosenbergs did not select the right attorneys.

The "Thou Shalt Nots" of Finding a Lawyer

I: Thou shalt not yield to the temptation to ask for a freebie.

I was sitting in the steam room after a work-out when in shuffled Ollie, plopping his rotund body next to mine. The sweat was dripping down my forehead and stinging my eyes. Another minute or two, I'd be cooked and ready for a cool shower.

"Mind if I ask you a quick question," Ollie screeched over a blast of fresh steam. Before I could answer that I did mind, and had no desire to field another legal "what if" and "suppose this and that" about some phantom friend, Ollie posed a fairly complex situation involving the collection of an overdue obligation from a debtor who filed bankruptcy. I had no qualm in giving him an answer worth the fee I received: I told him he should hire a thug to collect the money.

"There is no such thing as a free lunch." Then why expect one from, of all people, a lawyer? Time is money, and like most of us, attorneys are not about to part with it easily. As tempting as it may be to save money by using a lawyer who is a friend or relative, insist on paying. Even if the fee is discounted, payment insures that the services rendered will be professional.

The temptation to take advantage of free legal advice sometimes arises at the workplace, where you casually saunter into the office of the in-house corporate counsel and ask for the solution to some personal dilemma. The problem is, attorneys usually know very little about areas of the law outside their particular domain. Moreover, a potential conflict of interest would arise if the issue involves your employer.

The bottom line is, never go to a lawyer because you expect to pay no fee. Should a friend or relative offer to act for free, graciously turn it down.

II: Thou shalt not walk thy Fingers through the Yellow Pages.
Until the mid-seventies, listings of attorneys in the telephone directory were, like lawyers, rather boring, consisting of names, addresses, and phone numbers. However, this was useful information, and also a good source in selecting an attorney geographically desirable.

When prohibitions on legal advertising were declared invalid by the U.S. Supreme Court, lawyers, gingerly at first, entered the arena of promotion and public relations. Some restrictions still remain. For example, with the exception of patent, trademark, and admiralty law, forty-three states either ban or restrict advertising by attorneys of legal specialties, even when certified by recognized organizations. Hence, when you see bold block ads in red and blue on a field of yellow listing drunken driving or personal injury or real estate, and so on, this only means the lawyer or firm engages in those activities. It does not signify they are "experts" in a given field.

What is true for the yellow pages goes double for radio and television commercials.

"Hey, Jack, how's your back after the accident?" Bowler One asks.

"Not bad, Fred," Bowler Two answers, lifting his bowling ball with a grimace.

"You know, Jack, you really should see a lawyer. I got rear-ended once and was glad I had an attorney. Why, it didn't cost me a thing until the settlement check came through."

"Gee, Fred, maybe I should. Know one?"

Then the beaming face of Honest Abe Trustme, Esquire, permeates the screen with a phone number and the phrase, "No Fee Unless We Collect," flashing on and off. Meanwhile, back at the law offices of Trustme, Greed, and Avarice, ravenous lawyers, clutching contingency fee agreements, are waiting for the phones to ring off their hooks.

There is nothing wrong with advertising, but other than letting you know where a lawyer is located and what type of work the lawyer "supposedly" does, it is entirely worthless as a means of selecting the right attorney. Do not rely upon it.

III: Thou shalt not covet a law firm because it is large.

When it comes to attorneys, bigger is not always better. Since people tend to regard lawyers as "hired guns," there is the impression that the larger the law firm, the more advantageous for you, the client. But this isn't really so.

Although there are occasions when the size of the firm is a factor, cast aside any bias to go with the big firm. I have found small businesses often inclined to retain a large law firm in the belief that it will add respectability and lend prestige to their status. While this is true to some extent, any advantage gained is outweighed by the high cost and lack of attention paid the "inconsequential" client.

Here is an example of what can happen:

Only a year out of law school and having just started a firm with three other young attorneys, I signed up a medium-sized corporate client. My partners and I had no difficulty in fielding the legal matters the client generated, until one day when the client was served with a class-action complaint. Since class actions were fairly new at the time and the possible repercussions substantial, I thought it best to forward the matter to a larger and more established firm.

This is where I made my mistake. I had not yet learned the ways of the large law firms and, like so many people, thought bigger was better. As a result, my client paid dearly. The senior partner, to whom I was referred and who charged in one hour what I did in a day, had yet to learn that "class action" wasn't one word. As for the associate responsible for the bulk of the work—the alleged "expert" in this burgeoning area of litigation—I soon discovered he had, in fact, handled only one such case previously, although you would never guess this from his hourly rate.

After several months of filing motions, memorandums, and preliminary objections, all at very considerable billings, my client

cried "uncle" and my firm regained the case. I then arranged with
a lawyer at a small firm with an excellent reputation in civil
litigation to act as co-counsel. Together we handled the matter
and reached a satisfactory resolution. The fact that our total legal
fees were less than those charged by the large firm for only a few
months work was not lost on the client.

What it boils down to is this: Big firms want and serve big
clients. If you do not fit this category, you probably will encounter any or all of the following: unaffordable legal fees, work
performed on your case by underlings, a general attitude that
you are not terribly important, and the costs exceeding the gain
at the end. This is not to say that large law firms have no place,
but that place may not be for you.

IV: Thou shalt not worship political connections.
Bees are drawn to flowers; bears are attracted to honey; and
politics lures lawyers. They take to it like a fish takes to water.
Wonder why? One reason is because people have become convinced the way to get things done is to "know someone," "pull
strings," or use "politics". While this is certainly true in many
areas of our lives, it rarely applies to American jurisprudence.

I say this not out of blind loyalty to the law but because I am
tired of seeing attorneys pull this scam on the public over and
over again.

Unfortunately, attorneys are occasionally successful in
employing politics. Most of the time, however, lawyers are more
like illusionists projecting the appearance that they have been
able to get something done because they are politically wired or
"know" the judge, when in fact, whatever happened would have
happened anyway.

As for those instances when politics is effective, it is normally
limited to administrative and regulatory matters such as securing a variance from a zoning board or gaining a government
contract. Sometimes the sphere of political influence wielded by
attorneys spills into the private sector. For example, if you are
attempting to obtain a loan from a bank, perhaps you can acquire
a more favorable rate or a higher amount if the attorney negoti-

ating for you has ties with the local municipality which, in turn, has substantial deposits with the bank in question. In any event, the effects of political influence peddling are normally slight, except where there is outright corruption and bribery.

When it comes to matters involving litigation, it is especially important to retain an attorney for reasons of competence rather than political reputation. Consider this scenario taken from my early years of "innocence":

My client was a company doing business with the federal government. After providing the appropriate service, the company was denied payment amounting to several hundred thousand dollars. The available remedies consisted of administrative appeals or filing a lawsuit, which could only be brought after first obtaining permission from the government to sue in the first place. Thus, the prospect for a speedy resolution, critical to my client's cash-flow requirements, appeared bleak.

At the suggestion of another attorney who had been politically active for many years, my client paid a $5,000 retainer to a congressman who was also a partner in a law firm. I monitored the case while another attorney at the congressman's firm went through legal maneuvers amounting to nothing more than generating a sizeable bill. In reviewing the initial statement, I noted no credit was provided for payment of the retainer, and when I asked why, I was told the "retainer" paid to the congressman/attorney was separate and distinct from any fees due the law firm.

After a heart-to-heart talk with the client, I was given authority to terminate all dealings with the congressman and his firm. I demanded and obtained a credit for the retainer paid against the outstanding bills. I advised the government of my intention to proceed under the Federal Tort Claims Act and shortly thereafter we reached an agreement. Several years later, the congressman/attorney was convicted under an assortment of corruption charges and was subsequently disbarred.

At best, the legitimate activities of these "political" attorneys are rarely worthwhile. Unless you are willing to immerse yourself in the muck and mire of politics at its worst, I strongly urge you not to consider political reputation when hiring a lawyer.

How to Locate
an Attorney

In terms of raw data, the telephone directory is an excellent source for obtaining names, addresses, and phone numbers of lawyers. There is one set of books, however, which lists all the attorneys admitted to the practice of law in the United States. This tome is the *Martindale-Hubbell Law Directory.* Attorneys are listed geographically and alphabetically, with information including their age, date admitted to the bar, college and law school. There is also a limited rating system. *Martindale-Hubbell* is accessible in many libraries and all law libraries.

Once you have secured several names in this fashion, you may proceed to evaluate the candidates on the factors I suggest later in this chapter. However, you might wish to pursue one or more of the other avenues open to locate an attorney.

Lawyer Referral Services. Most local bar associations maintain a referral service which you can contact to obtain the name of an attorney. Sometimes these attorneys consent to an inexpensive initial consultation. Although the lawyers often indicate the type of cases they will handle, do not rely upon this as meaning the lawyer is an "expert" in the field (Thou Shalt Not II). Keep in mind that while referrals from bar associations may be adequate for an individual involved in a divorce or the purchase of a house, it will rarely do for the business concern in need of more specialized representation or for more complex matters.

Word of Mouth. An excellent way to find a lawyer is, like everything else, through word of mouth.

"Hey, Jack. I just played the best set of tennis ever because of my new racket."

"Yeah, Bill, what kind of racket?"

"Hey, Jack, the movie the wife and I saw last night was the greatest!"

"Yeah, Bill, which one?"

"Hey, Jack, let me tell you, this new after-hours club has got the most fantastic floor show."

"Yeah, Bill, where is it?"

"Bill, you look awful. What happened?"

"I don't know, Jack. Things aren't going so good. Ya know Beverly, the friend I been telling ya 'bout, the one I met at the after-hours club?"

"Yeah, Bill, the one that ..."

"Never mind that, Jack. Well ... the wife found out and after breaking my new tennis racket over my head, she threw me outta the house. I think I need a good lawyer. Know one?"

"Nah... sorry, Bill."

"Hey, Jack, the lawyer I got for my divorce is really something! Here, take her name if ya ever need one."

Now, you may not want Bill's advice concerning his tennis racket or what movie to see and certainly not where to go "after hours," but you should seriously consider taking the name of his lawyer for any domestic problems you may encounter.

Business and Professional Associates. In addition to friends and family, who can be an excellent source for learning the whereabouts of a good attorney, if you need representation in your business, other firms in the same industry or people in your line of work should be contacted.

Frequently, people turn to their accountants to recommend an attorney. This makes sense, but a word of caution is in order. I am often requested by clients to suggest an accountant, and it is unlikely I am the only lawyer in the world asked this question. So it is reasonable to assume that accountants are sometimes beholden to a particular attorney to return a favor. Therefore, obtain at least two names from your accountant and speak or meet with both.

Ask Another Attorney. As improbable as it may appear, asking another lawyer is an excellent way to find an attorney suited to your needs. Naturally, do not make this inquiry if the lawyer may be interested in having you as a client. However, there are many attorneys you can ask. If you know an attorney who does not do domestic relations work and you are involved in a divorce—ask. If your "family" lawyer is preparing to retire and is not accepting new clients—ask. If the lawyer who is "house counsel" at your company has no private practice—ask. If your business has an attorney on retainer who does not engage in the field in question—not only should you ask, but consider requesting that lawyer to monitor the case on your behalf, especially the billings.

Perhaps you have heard the expression, "a lawyer's lawyer"? These attorneys are the top of the legal profession since they are the ones other lawyers go to for help.

Needless to say, this lawyer would always be an outstanding choice and the only way you can learn of one is from other attorneys.

However, when requesting an attorney to suggest someone else, keep in mind that the payment of referral fees is a common practice. Therefore, determine whether the lawyer will receive compensation from the attorney recommended. If you are reluctant to ask, obtain two or more names.

Nevertheless, even with a paid referral, this is still a good way to find an attorney.

Whether you have obtained one name or five, and regardless of how—by phone book, by friend, by another lawyer, by a referral service, or by your accountant—the question of the "interview" arises.

Clearly, if you are a substantial client who will generate significant legal fees, or you have a claim involving a high monetary value, or you are in business with the prospect of a long, continuous relationship with your attorney, an interview for the sole purpose of deciding on a lawyer is reasonable to expect. Lawyers will be happy to accommodate you. Otherwise, forget it.

As far as ordinary matters are concerned such as buying or selling a house, defending a motor vehicle violation, having a will prepared, and so on, do not expect attorneys to make themselves available for only the possibility you may hire them. In such situations, schedule an appointment to discuss the case at hand. As the meeting progresses, based upon the considerations delineated below, decide if you want to hire that lawyer. If not, the most you will lose is a fee for initial consultation.

Factors In Selecting
Your Attorney

Feet come in all sizes and shapes. Basketball players have different sizes, all big, gigantic, or gargantuan.

Likewise, so it goes with the egos of lawyers.

Thus, don't bring a clipboard with a checklist and question-naire to the "interview". You won't even be asked to take a seat. Should the lawyer be willing to submit to this procedure, I think you have succeeded in locating a lawyer sorely in need of clients. This is not an attorney for you.

In a civilized (or as civilized as you can be with an attorney) manner, work into the conversation the factors I ask you to consider in hiring a lawyer. With some, such as whether you can afford the lawyer, you should be direct, while with others, like reputation in the community, you will have to feel your way through.

In any event, try to learn as much as possible so you can properly evaluate the attorney based upon the following criteria.

The Price Tag. I will never forget that moment, while out jogging along a fairly busy street, I was struck by the sight of it —a brand new, pale blue Mercedes 450 SL convertible. Over the years, I have had an unrequited love affair with that model of car, although sometimes the car wore red. Problem was, I couldn't afford one. Simple as that. And so it goes with lawyers. Like cars, there are some lawyers you just can't afford.

Although I never owned the Mercedes 450 SL, I did buy other cars, some of which were enjoyable and most of which got me to where I had to go. Similarly, there are affordable attorneys who can get the job done. Maybe not the Caddys or Jaguars of the legal profession but often a Chevy, or a Honda, or a Yugo (well, perhaps not a Yugo), will do.

How to deal with attorneys' fees will be explored in more detail in the next chapter, but you must determine at the outset what your case may cost. The best way to find out is to ask: "How will the fee be calculated?" and "How much is this going to run me by the time we are finished?", will get you off to a good start.

If the attorney is charging an hourly rate of $250 and you are involved in a divorce, you had best be prepared to spend many thousands of dollars. If this presents a financial strain, go to a lawyer who charges $100 an hour. Do not be afraid to admit you can't afford a particular lawyer, and whatever you do, don't delude yourself into believing the attorney will "cut you a break" or charge you less. Just move on until you locate a lawyer you can afford. Once you reach this point, it is time to consider the remaining criteria.

Experience. With the exception of specialties such as maritime or patent law, attorneys are not certified in specific areas. However, almost all lawyers specialize. Even the general practitioner who engages in real estate, business, and domestic work may never represent a defendant in a criminal proceeding. It is crucial, therefore, that you select an attorney with experience in the type of case you have.

Given the importance of this factor, it is best to be straightforward. Ask the lawyer what kind of work he or she performs. In a large firm, the area of concentration for any one attorney is likely to be narrow, although the firm taken together may handle all legal matters. However, even the largest law firms tend to specialize. Solo practitioners or members of small firms generally have experience in several fields of the law.

The more specific the issue or the more complex your case, the more critical it is that your attorney possess the requisite expertise. For example, if you want to incorporate your business, any lawyer who practices business law and has incorporated clients in the past will do. However, if you want to expand your three-store operation into a national franchising company, more than a business lawyer is required. Not only will you need an attorney specializing in commercial transactions but also one who is well versed in franchising laws.

While most attorneys are forthright in providing this information, you can verify the answer by inquiring about the attorney's client base. If you want to incorporate, ask the lawyer how many corporate clients he or she has. You may get some names, but don't expect permission to contact them as references unless you are very important to the lawyer. Attorneys' attitudes here are the same as they are toward interviews.

I was once asked for references by what could have been a major client. This occurred just after I left my former firm and struck out on my own, so I was especially vulnerable. However, after mulling over the matter, I politely declined to honor the request and I did not get the client. Like basketball players, I have big feet, size 12D to be precise. And like all lawyers....

Capability.

Question: What can turn a haughty attorney into an obsequious flunky?

Answer: A huge corporate client.

In the same way big businesses can command interviews, resumes and references from attorneys, they can ask almost any question they please. One question they often ask is whether the law firm has the capacity to handle efficiently and effectively the amount of work generated. If you anticipate you will require substantial service from your lawyers, this is an issue you must explore.

Obviously, the mega-firms have this capability but you should not automatically dismiss the medium-sized or even the small law firm. Be direct when inquiring. Determine how many lawyers and support staff will be available to you, and what other demands are being made on their time, such as lengthy trials or out-of-town commitments.

Choose the Lawyer Best
Suited for You and the Job

You arrive at the staid offices of Marcus & Welby for an appointment with Oliver Wendell Welby, Esquire. While you are waiting, you take note of the numerous plaques adorning the walls. Testimonials to the devoted services rendered by Attorneys Marcus and Welby from organizations including the Chamber of Commerce, Rotary, the United Way and the American Cancer Society. As pertains to O.W. Welby, there are diplomas from

Exeter Academy, Princeton, and Cornell Law School, as well as certificates of membership in several bar associations.

Precisely at the appointed time, you are ushered by a receptionist into the mahogany-paneled chambers of Mr. Welby who is sedately seated behind a meticulously maintained desk. He rises and extends his hand to greet you. "A pleasure, Ms. Client. Please be seated and Mrs. Stephens, please hold my calls," he pronounces. His handshake is firm.

By observation and without asking one question, you have come to know a great deal about Mr. Welby. The few minutes spent scanning the walls of the waiting area has revealed an excellent education and the high regard in which he is held by the community. You also have learned that Mr. Welby is punctual, courteous, and, most important, he will give you his full attention. Although first appearances and impressions can be deceptive, you must take advantage of this opportunity to learn as much as you can about the attorney you may hire.

"It's over half an hour since my appointment with Mr. Sleaze," you complain to the secretary seated only a few feet away in the cramped and stuffy reception room. "Oh... he's always late. Don't worry, he won't be long," she says.

Finally, the lawyer Sleaze emerges from his office, his arm around a frail, wobbly old man. "Jake, leave everything to me. Just remember, don't talk to no one, don't sign nothin', and keep mum. We'll make them pay for not scraping the ice off of their sidewalk!" Sleaze brandishes a cigar like a sword. The old man totters out. "Wait a minute, Jake," Sleaze calls into the hall. "Sandy," he barks, "Give the old geezer one of our canes and remind him never to leave home without it!"

So, do you hire Oliver Wendell Welby, Esquire, or retain the lawyer Sleaze? Welby, right? Not necessarily. The answer depends upon your requirements and expectations. Putting aside more objective standards such as expertise, which attorney is more affordable, capability to handle your work load, and so on, the time has come to chose the lawyer with whom you will mesh best and the one most likely to get you where you want to be. Welby or Sleaze? It all depends.

You are president of a small real estate brokerage firm and would like to have an attorney on retainer for advice and assistance as well as a lawyer who knows how to resolve and settle disputes out of court. You also envision this person will provide a valuable entree into the financial community. Welby or Sleaze?

In your mail one day, there is an anonymous letter disclosing that your business partner and wife have been "at it" for some time. You want him out of the company and her out of the house, and you don't care what it takes to do it. Welby or Sleaze?

For seventeen years, you have been nursemaid to your infirm uncle, which has cost you a career and social life. Uncle Amos has died and left you everything. The other nieces and nephews suddenly crawl out of the woodwork and seek to invalidate the will, alleging "undue influence." You want a lawyer who will reflect your integrity and who can make a respectable appearance before the jury. Welby or Sleaze?

Waiting it out at a red light, your car is struck from behind by a stretch limousine. Luckily, you are not hurt, but friends suggest you are entitled to compensation and should see a lawyer. Welby or ...

There you have it. The large law firm or solo practitioner? An attorney with an hourly rate of $300, one who charges $125, or one for free? A real estate attorney for a criminal matter, or a tax attorney for your divorce, or the general practitioner for a salvage claim under maritime law? An attorney arrived at by way of a fifteen-second TV spot or one obtained at the recommendation of your accountant and banker? No matter how, why, or whom you selected, you have now picked a lawyer.

CHAPTER FOUR

Ground Rules

Item: Attorney in New York mega-firm publicly concedes to having billed a twenty-seven hour day.

Item: Appeals Court overturns award of counsel fees decrying the "mutual back padding" of judges and lawyers when it comes to fees.

Item: In one nine-day period, an attorney billed a municipality for 154.5 hours (17.2 hours per day) and was paid! Critics have said he will "go down in history as the greatest lawyer."

With many law firms requiring their associates to submit 2,000 hours of billable time annually, fees have become a great concern to lawyers. Likewise, fees should be of even greater concern to you, the client. Therefore, the first and foremost ground rule is this:

Ground Rule I: Determine How You Will Be Billed

Like all ground rules, this must be established at the initial meeting. Beware the attorney who downplays the fee with a kindly smile and a nonchalant shrug. Would you hire a gardener or a plumber or a painter without knowing what you are going to pay? Same goes for lawyers. All reputable attorneys will discuss the fee arrangement and follow up with a letter of confirmation.

The way attorneys bill their clients, generally depends upon the nature of the case, and not the client's ability to pay. For example, while lawyers handle personal injury matters for a contingent fee, do not expect to find a competent attorney who will represent your business that way. Your cries of a bad season or a temporary adverse cash flow and promises of splitting what will be "millions" from a lawsuit are all destined to fall on deaf ears. Although there are hybrid fee arrangements, what you can expect will fall within one of the categories listed below.

The Hourly Rate. The hourly rate is the fee an attorney charges a client for one hour's worth of time. It is the mainstay of compensation in the legal profession and the most common method of billing. Sounds simple but it is not.

Not even lawyers always know how to determine their hourly rate. Frequently, it is an amount which seems in keeping with the competition, which translated, means they charge what the market will bear. I, on the other hand, recently raised my hourly rate to match that of my refrigerator repairman. Therefore, do not try to understand *why* a particular attorney charges what he or she does, just ascertain the amount and then whether it is "bundled" or "unbundled." The former is all-inclusive, while "unbundled" rates segregate additional charges for overhead and support (costs are always excluded and will be discussed separately).

Because most hourly rates are bundled, they may seem excessive, and like one international law attorney who bills $1,000 an hour, some are. But the fact is that up to half of what most attorneys charge goes for rent, office expenses, staff support, and automation. And to complicate matters with lawyers relying more and more upon computers and state of the art equipment, it is becoming increasingly difficult to set a fair hourly rate.

For instance, not many years ago it took a substantial amount of time to prepare and dictate a partnership agreement and an equally long time for the secretary to type it. Today, the attorney selects the appropriate boilerplate clauses stored in the word processor, and the secretary transposes it to print. Does this mean the attorney will charge less for the same product because of computerization? Either the hourly rate goes up to reflect the sophistication of office equipment or the attorney puts down more time than is actually spent.

If you are considering asking about the type of things you will be charged, save your breath. The answer is "everything." Whenever your attorney makes a move which in some way relates to your file, a time slip will be prepared. This may or may not include the interlude spent pondering your case while doing something else, a not uncommon practice among new associates in the mega-firms, given the enormous pressure they face to generate billable hours.

One thing you can do to prevent inflated legal bills is to keep a log and record each time you speak or meet with your lawyer and for how long. This way your five-minute phone call won't become seven minutes and kick into an extra billable unit, which in turn can cost an additional fifteen to fifty dollars. Moreover, the lawyer's conformance with your logbook is a good barometer of the accuracy of the other time slips.

You must also know the hourly rate and the increments in which it will be applied. Most firms charge in segments of one-tenth of an hour (six minutes) but others may utilize fractions of 1/6 or even 1/4. So a "two minute" phone call to your lawyer will be reflected by a time slip of six, ten, or fifteen minutes, and depending upon the hourly rate, can cost you up to seventy-five dollars. Although some attorneys won't charge for a "two minute" call, most will, unless you make this one of your ground rules.

Another reason to ascertain the unit of time in calculating the bill is that the hourly rate may be deceptive. For example, a lawyer who charges $200 an hour is more expensive than one who bills at $160. Right? Not if the first lawyer employs six-minute units and the second uses one-quarter hour segments. Your two-minute phone call to the first attorney will cost you $20. For the same call to the second "cheaper" attorney, you wind up paying $40.

It is possible more than one attorney may work on your case and, if so, find out his or her identity, experience, and hourly rate. Beware the senior partner's "it's to your advantage" smile when he explains he will reserve his $300-an-hour services for the important issues while the $145-an-hour associate will do the research, draft the pleadings, and so forth. What you will not be told is that it may take this recent law-school graduate three times as long to do the same thing, or that the senior partner may spend significant time (for which you are charged) in reviewing and correcting the subordinate's work. On the other hand, if the associate is experienced, this can be to your benefit.

When two or more lawyers are assigned to your case, you had best be vigilant of the practice known as "double billing". Once, on behalf of a client, I attended a conference of all defense counsel in a multi-litigant lawsuit. One of the defendants was represented by a senior partner and an associate from a major law firm. Although very little was accomplished, a good deal of time was spent accomplishing it, for which the client with the two attorneys was billed over one thousand dollars—both filed time slips. To add insult to injury, the associate billed the client for the half hour we all waited for the other member of the duo to show up.

It may become necessary for you to keep watch over a herd of lawyers. For example, in the "item" at the beginning of this chapter where the Appeals Court overturned an award of legal fees, the court was troubled that the firm submitted time slips for five partners, six associates, two summer associates, and seven paralegals - all over a straightforward real estate transaction with a purchase price of $340,000!

While there may be occasions when it is appropriate for two or more lawyers to bill the client concurrently, such as during the course of a trial, your permission should always be required. Furthermore, never consent to paying double (or triple) for the infamous "inter-office conference".

Sometimes this is no more than legal jargon for shooting the bull about a case. Even if beneficial, why should you pay for one attorney to serve as a sounding board for another attorney you are paying in the first place!

The hourly rate, if applied correctly and in accord with the proper ground rules, almost always results in an equitable way for a lawyer to earn a fee. Any type of case can be billed in this fashion, but there are other methods more typical of certain matters.

Fixed Fee. The "fixed" or "contract" fee is a flat charge for a specific service. This can apply to a variety of matters, including the preparation of wills, trademarks and patents, setting up a corporation, or filing a fictitious name registration. Sometimes criminal cases are billed this way as well. Contract fees are occasionally employed in domestic disputes, but here you must be wary of the quality and scope of services. Take the case of little Ms. Muffitt.

Lawyer: That's right, Ms. Muffitt, two hundred dollars, plus costs, for your divorce. No problem.

MM: Oh that's wonderful, Mr. Spydier, because I really can't afford more.

Lawyer: Fine. Now let's get started. Tell me my dear, why do you want a divorce?

MM: Because my husband beats me (says little Ms. Muffitt, meekly alluding to a severely bruised tuffitt).

Lawyer: Why that's terrible! We won't stand for that! I'll get a protection-from-abuse order.

MM: Is that included in the two hundred dollars?

Lawyer: No, but we can't allow him to go on hitting you, can we? Ok then, let's move on. How much do you earn a week?

MM: I don't work, Mr. Spydier. I'm busy caring for the twins. They're only three.

Lawyer: Oh, in that case we will also have to apply for support and alimony.

MM: Will this be extra, er ... your fee ...

Lawyer: Yes, but don't worry. I'll file a petition for your husband to pay counsel fees.

MM: This petition will be an additional charge, won't it, Mr. Spydier? (Ms. Muffitt's blue eyes narrow with a look of suspicion. She's beginning to catch on.)

Lawyer: Correct. (less friendly, more defensive, like a kid getting caught with his hand in the cookie jar).

MM: I don't know, Mr. Spydier. I was hoping this would be simple and cost only two hundred dollars just like your ad said. All I want is the house and ...

Lawyer: Oh, there is property involved? Well, that complicates matters a bit .... Now this can be hard to estimate.... H'm... I think I'll have to apply my hourly rate of ...

In other words, the cheap contract fee was a typical come-on, to be used as a "bait and switch" for the upgraded model. The advertised fee was good only for a divorce with no joint property, no children, no question of support, and nothing required other than the shuffling and filing of the necessary legal papers. While the fixed fee is a legitimate method of establishing attorneys' charges, be very clear at the outset what it does and does not include.

Contingent Fee. In a contingent fee arrangement, the risk is shared between attorney and client. The lawyer gets paid only for a successful result, and the fee is a percentage of the amount won.

If clients had their way, attorneys would almost always be paid, or not paid, in this fashion. However, lawyers take cases on a contingency only when they have a good expectation for recovery. Customarily, contingent fee arrangements are reserved for the personal injury case, debt collection, and occasionally insurance claims against one's own insurance company.

Personal injury lawyers may leave home without their American Express card but they will never be found without a contingent-fee agreement at their fingertips. This document is straightforward enough but there are two things you should be very clear about.

First, although you must authorize any settlement of the case, some agreements provide that if you refuse to approve a settlement recommended by the attorney and the case goes to trial, you must pay the attorney's fee based upon an hourly rate, should you lose. Second, and most critical, is the manner in which the lawyer's percentage is calculated.

The attorney's "take" can vary between 25% and 50%, depending upon the risk factor assessed by the attorney and when the case is consummated. The lawyer's share increases at different stages. For example, the lawyer receives 25% if recovery is reached before suit is filed; 33% between the commencement of an action but prior to the trial beginning; and 40% thereafter. This is fair, since the attorney is entitled to more if greater effort is expended.

However, sometimes cases are settled at the courthouse steps on the day the trial is scheduled to begin. In order to prevent your attorney from taking an extra seven or ten percent for a few hours work, you should have the contingent fee agreement specify that the final percentage is reached only if the case is actually tried.

The main thing is not the fraction involved, which is clearly stated, but whether it will be applied to the gross amount obtained or the net sum, which takes expenses into account. It is always to your advantage that it be the latter.

Take a case with a recovery of $10,000. The attorney is entitled to forty percent, but there are $2,000 in expenses, which include costs of transcripts, expert witness fees, unpaid medical bills, etc. If the lawyer's percentage is off the top, the fee is $4,000, the same you receive, making it a 50/50 split. On the other hand, if calculated as a percent of the net, the attorney is paid $3200 and $4800 goes to you, a true 60/40 division, which is what you had in mind at the beginning.

Percentage Fee. Like the contingent fee, this method of compensation is calculated as a percent, but here the payment is absolute and due at a certain time. The fee is tied to the value of the transaction as well as the anticipated amount of work involved. Percentage fees are utilized in matters such as estates, bond issues, and sometimes real estate closings.

When applied to estates, the percentage is reduced as the size of the estate increases. This makes sense because often the same amount of effort is spent on the small estate as the large one. On the other hand, for real estate transactions, the fee is normally 1% of the sale price, and downward adjustments do not take place. The percentage fee arrangement for real estate is nothing more than an attempt to snatch an additional piece of a rather expensive pie (see "value billing" below).

The fact is that it is not unusual for more complications to arise in the sale of a $100,000 house with the buyer attempting to secure an FHA mortgage than in the sale of a commercial property for $1 million to a cash purchaser. This is why lawyers apply an hourly rate if the price is not substantial. On real estate matters, pay the attorney an hourly rate, subject to the ground rules, and you will be ahead.

Value Billing. Almost everyone has heard Abraham Lincoln's "lawyer's creed," that "a lawyer's time and advice is his stock and trade," but the world has taken little note, nor long remembered, the "other" lawyer's creed formulated by the distinguished Professor Leonard Leech of Cornell Law School (or so we are told by his creator, Kurt Vonnegut, Jr.). Yet, I have found that Leech's "lawyer's creed," perhaps unknowingly, has been taken to heart by every attorney I have met. Here it is:

In every big transaction, there is a magic moment during which a man has surrendered a treasure, and during which the man who is due to receive it has not yet done so. An alert lawyer will make that moment his own, possessing the treasure for a magic microsecond, taking a little of it, passing it on. If the man who is to receive the treasure is unused to wealth, has an inferiority complex and shapeless feelings of guilt, as most people do, the lawyer can often take as much as half the bundle, and still receive the recipient's blubbering thanks.

This is the theoretical foundation upon which the system known as "value billing" rests. It is the attempt, almost always after the fact, to set a fee based upon the "value" of services rendered. This should be distinguished from a bonus which is discussed and agreed upon in advance.

When I was still a tyke, I palled around with two very good friends, Joey and Stevie. From time to time, one or the other would have a nickel and buy a candy bar, usually a Three Musketeers. In those days, there were two chocolaty raised ridges demarcating three equal sections. The way I figured it, why should Joey or Stevie have the entire bar of candy when it was a simple matter to break it into three pieces, one for each of us. Once I convinced my friends of this, I was never without my allotted share of a Three Musketeers bar.

This is "value billing." It occurs when a voracious attorney craves a chunk of your candy bar. Don't be cajoled into parting with it. When you hear "value billing," hold on to both your chocolate bar and your wallet!

Hybrid Billings. Given the growing use of computers, the increasing complexity of legal issues, and a greater sophistication of the financial structure of law firms, there is a tendency to adopt a more flexible approach when invoicing clients. Therefore, you may encounter a fee arrangement which is a combination of those listed above, or perhaps none of the above. Some you could come across are:

The Adjusted Hourly: The client is billed an hourly rate which will be adjusted to reflect the difficulty of the task and the result obtained. Thus if a favorable outcome is achieved, you may be charged 150% of the standard hourly rate, but only 75% of the rate if the result is unsuccessful.

Hourly/Contingent: In exchange for a discounted hourly rate, the attorney will keep a percentage of any money acquired as a result of his or her actions.

Hourly/Fixed: This is used in two ways. In one, a contract price is set, but if more work becomes necessary than originally anticipated, the hourly rate kicks in. The other limits the total arrived at by the hourly rate: it cannot exceed a preestablished fee.

How Much Have Ya Got? "Res Ipsa Loquitor" is a legal maxim which translated from the Latin means, "The thing speaks for itself." So does this fee arrangement. It is the ultimate hybrid, and an attorney using it will bastardize all other legitimate methods to separate you from as much of your money as possible.

I have spent a good deal of time on the matter of fees because it is the most important ground rule. Despite an attorney's best intentions, there is a conflict of interest inherent in the system. The client doesn't want to pay more than is absolutely necessary, and the lawyer has a business to operate. Therefore, without knowing how you are going to be billed, you will never know whether the attorney is doing it fairly and correctly.

Before moving on, there are a few remaining items to clarify. Determine when payment is expected and if there will be finance charges on unpaid balances. If you pay a retainer, make certain you know how it will be applied to your bill. Remember the case of the "vanishing retainer" involving the congressman/attorney?

Finally, without exception, the fee arrangement should either be reduced to an agreement or confirmed in the engagement letter you receive from your attorney. Once this is accomplished, you are in position to establish the remaining ground rules.

Ground Rule II:
What About Costs?

"Cost" is defined as the "outlay or expenditure of money, time, trouble, etc." On the other hand, "profit" occurs where "financial gain" results. Seemingly, the two are not synonymous. Explain this to lawyers.

I knew of a firm which billed its clients twenty-five cents per page for photocopying. In turn, the firm contracted with an outside company at ten cents per page. Subsequently, its appetite barely whetted, a new company charging only six cents per page was hired. The results, an astounding profit to the law firm of $40,000 in one month!

Control of costs is an important ground rule. Do not be proud. Whatever you do, never give your lawyer the impression you have a deep pocket. Find out what your copy charges will be. If it appears excessive, make it clear you want the option to do the copying yourself. Where a large volume is involved, try to negotiate a discount. After all, the major clients do.

Lawyers love to act important, so they have a flair for sending documents by overnight mail or courier service. With the arrival of the "fax" machine, letters are transmitted in a matter of seconds. Most of the time, this is unnecessary, since at the receiving lawyer's end the item is placed on a pile where it sits for days or weeks, after which the response is similarly shipped or faxed. Meantime, guess who pays the costs?

A special word about contingent-fee cases. Although your attorney is not entitled to a fee if there is no recovery, this does not hold true for the costs which the lawyer has paid in advance. Though you receive no money should you lose, you will receive a bill from your lawyer for reimbursement. Therefore, as the case moves along, keep an eye on expenses. One way or the other, you will pay for them.

While you cannot require your lawyer to obtain your permission for placing postage on every letter, your approval should be sought for any significant expense for which you will be charged. By 'significant," I mean any amount which is significant to you. If you grant your attorney unbridled discretion, you may discover that you have been "faxed" into poverty.

Ground Rule III: Make
Your Expectations Known

Often, especially in domestic matters, clients express uncertainty, confusion, and ambivalence regarding their wishes. So, as best you can, be crystal clear and let your lawyer know what you want, or the results can be counter-productive.

For example, let us assume one of your customers has run up a large, overdue account and your phone calls and letters have been ignored. You are left with no alternative but to refer the matter to an attorney, yet you would like to leave the door open to resume future business. If you don't make your attorney aware of this, it is probable the first letter sent to your customer will destroy any potential for further dealings. Therefore, tell your lawyer there is more involved than collecting the debt, and make certain he or she can reflect these sentiments in the negotiations. This brings up the importance of establishing rapport with your lawyer.

Not only should your personalities dovetail, but the lawyer's expectations must conform with yours. This is impossible unless you are both starting from the same place. For instance, if you are convinced the merchandise delivered was faulty and therefore you are under no obligation to pay for it, but your attorney isn't so certain, resolve this issue before hiring the lawyer. It may take some time and your lawyer may have to do some preliminary work, but wait until you both are seeking the same result before going further.

In order to accomplish this, it is necessary that you and your attorney be honest with each other. At the initial meeting do not tell your lawyer you are prepared to settle a claim by paying fifty cents on the dollar if you can't come up with the money. Nor should you keep information from your lawyer, no matter how damaging or embarrassing it may be.

The fact that you propositioned your secretary *before* firing her will definitely be brought to light during the course of a wrongful discharge action. Be candid and reveal everything to your lawyer concerning the case.

In turn, you have every right to expect your lawyer to be up-front with you. Salespeople are prone to exaggerate the benefits of their product. Within reasonable bounds, this practice is known as "puffing," and is not illegal nor fraudulent. An attorney eager to bag you as a client may likewise "puff" a bit. To an extent this is tolerable, but be on guard of the lawyer who boasts an ability to get everything you want.

Ground Rule IV:
Find Out Who Does What

If your lawyer plays Bud Abbott to your Lou Costello and the explanation about what the partners, associates, and paralegals do sounds a bit like "Who's on first, What's on second, and I Don't Know on third," call time out.

In medium-size and large law firms, often more than one attorney will work on your case. You will probably face a combination of senior and junior partners, associates, and paralegals, each of whom has a separate hourly rate. The effect of this on the fee has previously been discussed, but it also can affect the quality of the representation you receive.

There is always a lead attorney in charge. It may not be the lawyer you first hired if it is subsequently determined that another member of the firm has a greater expertise, or the nature of the case has changed. For example, you may retain one lawyer to persuade the township to look favorably upon your application for a zoning variance, but once denied and with the case headed for litigation, the chief trial lawyer of the firm may take over.

You have the right to determine whether the attorneys assigned to your case are acceptable. Therefore, make it clear you wish to be consulted and your approval obtained before someone is added to the team. You should base your evaluation of this individual on what you learned in Chapter Three.

Sometimes it is not a question of who works on your case but whether anyone in the firm should be performing a task for which you are billed. Once as officer and general counsel of a corporation, I referred a matter to a law firm which then delegated to a paralegal the chore of correlating all relevant correspondence and documents. This consisted of several file cabinet drawers crammed with letters, reports, agreements, and the like. For her services we were billed $55 per hour. When I became aware of this, I ordered the project halted and had one of our secretaries continue to categorize the material.

Or take the case of the doctor who was too busy to find out when his daughter, a witness in his support hearing, would be arriving from college. Instead of checking the train schedule and coordinating this with the hearing, he left it to his team of lawyers who billed over three hundred dollars for their time, including an inter-office conference to decide whether to book the Metroliner or the local!

This illustrates another aspect of this ground rule, which is to decide what your lawyer will do and what you can do just as well. Remember, if you are paying by the hour, your attorney will be compensated for whatever work is undertaken at that rate structure. So should difficulty be encountered in locating the whereabouts of a defendant and your lawyer spends two hours at $200 an hour contacting the post office, driving to the registered corporate headquarters to confirm occupancy, and searching through telephone directories, you pay $400 for detective work. On the other hand, if you do it yourself, you save the fee.

Incorporate into Ground Rule IV the option that you or your staff will have the opportunity to perform the nonlegal work. Since this is hard to predict in advance, you can only make this determination if you are aware of just what is going on with your case. Now we arrive at the next, and last, ground rule.

Ground Rule V: Be
Responsive and Stay Informed
You are ushered into the stylish office of Patricia Pompas,

Esquire, your prospective attorney regarding the acquisition of a business. She comes highly recommended by your accountant and banker. Ms. Pompas has an excellent reputation for being a deal maker, and commercial law is her *forte*. So far, you have done everything right.

Ms. Pompas motions for you to sit while she concludes a telephone conversation. With little else to do, your eyes roam the terrain of her desk: several legal pads, an appointment book with few empty slots, pencil holder, pen set, dictaphone, an expandable folder containing a half dozen files. The rest of the desk is cluttered with a score or more of pink telephone messages assorted in tiny clusters of threes and fours. Must be an important lady, you think. She's for me. You have just made your first mistake. Chances are, if you retain Ms. Pompas, in a few weeks one of those pink slips will be your unanswered telephone message.

The most frequent complaint leveled against lawyers is their failure or tardiness to return phone calls. Be emphatic that you want your messages promptly answered, and provide an evening phone number in the event your lawyer is in court during the day. One of my former partners considered a call returned if he dialed the number. The fact the line was busy or no one answered was irrelevant, so you had best define what is meant by a "return call".

You must also be responsive. While your case is pending, your lawyer may be in immediate need of information which you alone possess, or a decision must be reached whether to agree to certain terms in order to complete a settlement. Responsiveness is a two-way street and can be attained only if both parties respect this ground rule.

The adage "Ignorance is Bliss" may work well in some situations, say for certain former presidents asked to recall their involvement in the Iran-Contra Affair, but it is a major handicap to you when working with your attorney. Insist on being kept informed.

The best way to accomplish this is to require that you receive copies of all correspondence and material pertinent to your case. Not only will this enable you to monitor its progress and make intelligent decisions, but you will have the information you need to evaluate the accuracy of your bills.

Speaking of bills from lawyers, it was my responsibility to approve the invoices issued by the firm with the $55 an hour paralegal. After several months, I noticed each bill included a charge for "file memos" reflected by a 2/6 hour time unit and an hourly rate of $175. Fairly expensive, with one or two such memos per statement. Had I insisted, which I confess I did not, on receiving copies of these memos, I would have been in a better position to determine if the bill should be approved.

Citing myself as an example of what *not* to do, let me point out what is sometimes the most difficult part in handling your lawyer. In the above instance, not only was I the referring attorney but the client as well. Yet I did not insist on receiving copies of the file memos. Although the ultimate result was successful, the law firm was expensive and sometimes unresponsive during the course of the matter. Why didn't I do something about it?

I think it is because we give up some of our independence and critical faculties when we place ourselves in the hands of those whose help we seek, such as the physician or teacher or therapist. So it goes with lawyers. And when we do this, in some ways we mirror the behavior of children relating to their parents. Respect becomes confused with submission and alliance with acquiescence. We hesitate to speak our minds. We forget who is working for whom.

I have seen this over and over, time and again in people who complain to me about their attorneys, in clients whom I have referred to other lawyers, in the way my clients behave to me, and in myself. Therefore as easy as it may sound, the No. 1 Rule in this book is: You are the boss!

CHAPTER FIVE

You're the Boss!

Remember, you are the boss. This does not mean you should overrule your lawyer on matters of law like Mr. Pto Mein did when he opened a Chinese restaurant.

On the advice of his lawyer, the business was incorporated in order to insulate Mr. Mein from any personal liability. The lawyer made it a point to caution his client that he should sign all documents as president of the company and not individually.

Unfortunately, things began to sour and the business was in need of cash. Contrary to counsel's strenuous objection, Mr. Mein personally guaranteed notes with the bank, purchase orders from the food suppliers, and even the lease.

Despite an admirable effort to keep the business afloat, the ship sank and like a dutiful captain, Mr. Mein and all his personal assets went down with it. Had Mr. Mein heeded his attorney, his losses would have been limited to the initial investment.

Lawyers are learned in the law and trained to think logically. Therefore, listen to your lawyer and weigh the advice you get, but remember that the final decision is yours.

Obviously you will not endear yourself to your attorney by second-guessing the legal issues. In addition, people have a tendency to become obsessed with their case at the expense of every other aspect of their lives. For instance, it is not uncommon for business people to devote so much attention and time to a lawsuit that in the end, regardless of the result, their business either suffers or fails. There will always be enough for the client to do during the course of litigation, so leave the law work to the lawyer and stick to doing what you do best—running your business.

Think of it like delegating any task or responsibility. Do not get bogged down in the nitty-gritty. Respect the expertise of those in whom you have placed your confidence. Yet maintain an overview, check up on a regular basis, and be the boss.

Just when to assert your authority over your employees or subordinates is not always clear. The same holds true for knowing when to do this with your lawyer. This is not an exact science, but there are some criteria, and the examples that follow should give you a feel for the right time to throw your weight around.

Keep Your Lawyer's Nose
Out of Your Business

Once upon a time, there lived an enterprising young attorney named Jack. One day, a farmer named Duckinbill made an appointment to see Attorney Jack after normal hours. At this clandestine meeting, farmer Duckinbill furtively disclosed he had come upon a goose that was no ordinary goose.

"How so?" Jack asked.

"This goose lays golden eggs," whispered farmer Duckinbill.

Naturally, Attorney Jack, being a lawyer, was skeptical. But following a demonstration and two golden eggs paid as a retainer, Attorney Jack agreed to represent farmer Duckinbill in the sale of the goose.

After a while had passed and various offers had been received, the giant Levi A. Thon emerged as the high bidder. Attorney Jack spent many hours in negotiations with the giant to hammer out the terms of an agreement. The assiduous attorney wanted to cover all his bases and obtain the best price possible (and, no doubt, pick up a few more golden eggs for his time, since his fee was set at one egg per hour).

Weeks went by as discussions continued on such issues as warranties to be extended by farmer Duckinbill, reversionary rights to goslings born to the goose, collateral for the additional payments due after closing (the giant's castle was mortgaged to the hilt), and so on. Just as negotiations neared completion, and while in the process of dropping an exceptionally large egg, the goose suffered cardiac arrest and died. No goose, no golden eggs, no deal.

Word quickly spread far and wide of the misfortune suffered by farmer Duckinbill. As a result, deserved or not, Attorney Jack saw nary a client again, and from that day forward, he ignominiously bore the title of "Jack the giant (deal) killer."

Farmer Duckinbill had made the mistake of conferring unbridled discretion on his lawyer. Today, Attorney Jack's legacy lives on in the image of lawyers dubbed "deal killers." For example, consider the following.

Recently, I was involved in the sale of a thriving corporation from the sole stockholder to his two managers. The managers knew the business as well as the owner, so negotiations were conducted with everything on the table. The price and terms were agreed upon, with the seller providing financing and permitting the buyers to borrow the down payment from the bank.

As attorney for the company and personal lawyer for the seller, I handled the transaction. When asked by the buyers to represent them as well, I declined, explaining this would be a conflict of interest. Instead, I suggested they retain another attorney just to "hold their hands" through the contracts and closing. The managers followed my advice which, unfortunately, almost proved to be their undoing.

Despite being informed by his clients that this was a "done deal", the attorney for the buyers tried to renegotiate the sale price and terms, draft a completely different contract, and even went so far to suggest that the entire purchase was a mistake. To top it off, his fee was four times the amount I ultimately billed my client.

With the seller threatening to withdraw, the buyers decided to proceed without counsel and asked that I represent them in negotiating the loan with the bank. I agreed, and shortly thereafter we closed.

I continued to represent the company. My first assignment was to resolve the outstanding bill with the other lawyer, which was settled at less than one-quarter of the original quote.

This is not to say it is wise to engage in such matters without counsel. However, don't let your lawyer kill a deal for you. Either pull in the reins or hire another attorney.

To Settle or
Not to Settle

That is the question, and don't let your lawyer answer it for you. Despite the fact that attorneys are duty bound to represent the best interests of their clients, this is often easier said than done.

On the one hand, lawyers try to follow Honest Abe Lincoln's admonition to 'discourage litigation,'' and point out to the client that "the nominal winner is often a real loser—in fees, expenses, and waste of time." Yet the more protracted the matter becomes, the more substantial the flow of dollars to the lawyer. Therefore, only you, the client, should make the decision on when to settle.

What factors do you consider? First is your lawyer's advice. If you followed Chapter Three and hired the right attorney, you will get a sound professional opinion. However, analyze the reasons your attorney gives, and balance them with other factors. Such as?

For one, as previously mentioned, litigation can be very demanding of your time. It can involve depositions, supplying great quantities of documents, all of which you are convinced are highly irrelevant, the constant back and forth with your lawyer who will require information as the case progresses, and so on. I once represented a company where the request for production of documents was so extensive that for several weeks business came to a virtual standstill while employees from clerk to president sifted through cabinets and cabinets of files. You will be tied up by a lawsuit. Count on it.

Another point to keep in mind as you ponder what to do is the need to be pragmatic. What of your reputation in the business community? Would you prefer being known for having a hair trigger on your legal gun or as someone inclined to be reasonable and open to compromise? What about the people involved? Will you deal with them again? See them again? Do they mean anything to you?

There is much to evaluate before deciding whether to settle a legal dispute, but it should be obvious that *you* must make the final decision. Several examples will illustrate the point.

After twenty-one years of marriage, you and your spouse agree to put your marriage to bed without either of you in it. After all, the kids are grown, you each have independent careers, ownership of the house is clear and there is money in the bank. Why stay together when neither of you loves the other any longer? You agree, and begin to make arrangements for an amicable divorce.

Then the bombshell. You discover your wife has been having an affair and all this talk about "incompatibility" and "shouldn't there be more," is just a sham for wanting to run off with her new beau. Now you're pissed. You hire the nastiest, meanest domestic relations lawyer you can find. With vengeance in your eyes, you gladly fork over the $10,000 retainer.

Things get downright bloodthirsty: meetings, conferences, hearings, each a battleground for blood to be let. Petitions for support, alimony, and counsel fees are rattled like sabers. Bank accounts are raided by whoever gets there first. Other assets are frozen. Your kids are torn between two armed camps. You are losing sleep, your job suffers, your blood pressure is sky high, and your monthly statements from Attorney Hunn are always in the four figures (the retainer was devoured in the first three months).

You begin to reconsider. Your anger has abated. Anyway, you have found happiness with someone else. Time to back off. You call your spouse and schedule a truce. Things are looking up. Now to curb your bellicose attorney.

Settle?! Attorney Hunn will hear none of this. She bombards you with arguments about why you must persevere. "Forget the costs! There's much more at stake!" she bellows. Let her handle everything. Don't cave in, she enjoins. Settle or not? Listen to your lawyer? Settle.

When it comes to contingent fee cases, it would seem the interests of you and your attorney would have to be in harmony. If you get nothing, the lawyer gets nothing. The more you get, the more the lawyer gets. Thirty percent of a million is more than thirty percent of a thousand. Right? Not always.

Remember Ms. Prudence Juris, Esquire, in Chapter One?

She urged you to settle your lawsuit even though it would have been worth more later because she had a cash-flow problem. On the other hand, the reverse can be true. It may be you who can really use the ten or twenty thousand dollars right away, and your attorney would just as soon wait to settle for a higher amount, or take a chance on trying the case. Your lawyer has more than enough revenue this year, and wants to defer additional income for tax reasons.

Or your case could be one of a group where a settlement can generate a huge legal fee although the money going to each claimant is not great. For example, you may be one of fifty asbestos victims and your lawyer receives an offer to settle at $15,000 per case. He has a one-third contingent fee agreement. End result is he gets $250,000, while each plaintiff gets $10,000.

Then there is the doctor who was served with a complaint filed on behalf of his former spouse from whom he had been divorced for ten years. His Ex was seeking reimbursement for sundry expenses totalling $18,000 she had unilaterally spent over the years on behalf of the children. As his friend, I tried to mollify the outraged doctor, suggesting that things had been fairly amiable until now, so why not offer to pay half and be done with it.

The doctor turned a deaf ear. He engaged a heavy hitter in a big firm and paid a $5,000 retainer. There were some preliminary skirmishes, and finally a one-day hearing at which the judge rammed a settlement down the sparring ex-spouses respective throats.

The bottom line? The doctor paid $12,000 of the amount at stake. His retainer was $5,000. Then he received the final bill for the balance due his lawyers (a team at $250 and $175 an hour respectively) just shy of $11,000. Grand total - $28,000. As for his ex, her legal fees were more than $7,000. Which brings me to Bank's Rule of Thumb for Knowing When to Settle.

If the potential legal fees exceed the amount in contention, settle out of court.

As you can see, the decision to settle a case must be yours. Carefully consider all the factors, including your lawyer's recommendation. Then inform your attorney of your decision. Let your lawyer know you are the boss.

There are times when your lawyer will tell you how to behave or what to do. Attorneys take to heart the self-designated title of "counselor", and are forever offering advice. When do you follow it and when do you not? In other words....

Do I or Don't I
Listen to My Lawyer?

Whenever this question arises, I relate the following anecdote.

Seated across the settlement table from the buyers and their attorney was a builder who had developed hundreds of single family houses. Mrs. Buyer mentioned in passing that the front bricks were a shade off the color of those on the sample model. The builder matter-of-factly pointed out that this variance was permitted in the Agreement of Sale so long as the quality remained the same. That appeared to be the end of it, until their lawyer awoke from his daydream.

He went crazy, flying into a rage, ranting and raving, making a venomous attack on the stone-faced builder. For whatever reason, he was obviously having a bad day.

In such situations, the builder had a simple policy: politely smile, pack up the brief case, and walk away—all of which he did in a matter of seconds. As he departed, Mrs. Buyer burst into tears. She loved the house and wanted to move in. Her husband tried reassuring her, though not too convincingly, that their lawyer knew what he was doing.

Although not obligated, the builder returned the deposit money, and within several weeks resold the house at a higher price. A few months later, the same buyers, rather sheepishly, inquired if there were any remaining lots in the development. There were none. Mr. and Mrs. Buyer lost their dream home because of their lawyer. It was as simple as that.

Do you or don't you defer to your lawyer? The rule is this: If the decision turns on a legal question, comply with your lawyer's advice. Otherwise, you are the boss.

Chances are your attorney will be the one to tell you when this is so. After explaining the law, analyzing the alternatives, and making a recommendation, an attorney should end by saying, "But the decision is yours." On those occasions when there is but one course of conduct given the law, the lawyer may add, "If you don't do this, you better retain another attorney." Unless you hear these words or something to that effect, never feel compelled to do what the lawyer suggests. You decide.

Stand By Your
Convictions

I recently was asked for my opinion about the following incident by a gentleman whom we shall call Hugh M. Bean. One day, Hugh was driving along Collins Avenue in Miami Beach when, apparently out of nowhere, a wobbly old man stepped into the path of his car. By swerving the vehicle, Hugh managed to merely graze the leg of the straying pedestrian.

The old man was checked out at the hospital and then taken home. A police report was filed. Hugh called his attorney, who issued the standard instructions, "Do nothing and speak to no one. Refer everything to me."

Although the accident was only partially his fault, Hugh felt very bad. He bought a cake and brought it to the old man's apartment. He did some shopping for the old man the next day. They chatted a bit. He may have said something about being sorry the accident happened. Two weeks later he was sued.

Did he do the wrong thing by not heeding his lawyer, Hugh wanted to know. I asked Hugh how he felt about what he did. He said he felt good. Then he did nothing wrong, I assured him. He was just being Hugh M. Bean.

Which brings me to the story about Henry David Thoreau and Ralph Waldo Emerson. It happened at a time when Thoreau was in jail because he refused to pay taxes. Spotting Emerson walking in the commons, Thoreau shouted hello. Surprised to see his friend in a jail cell, Emerson cried out, "Henry, what are you doing in there?" Without an instant's hesitation, Thoreau called back, "Ralph, what are you doing out there?"

Regardless how you feel about civil disobedience, I think we can all agree there is a higher authority than the laws enacted by society. Whether we call this conscience or religion or morality or ethics, it is something about which the law is blind and lawyers are taught to ignore. Here you reign supreme. You are the absolute monarch and your lawyer is but a mere subject. Consider the following scenarios.

You are the third-generation president of a family-owned and operated concern which manufactures and distributes widgets. Some of your customers have been doing business with the company for decades. They rely on your integrity and take you for a person of your word.

At one of the weekly meetings with your quality control team, you are informed that the widgets shipped during the last three months are not consistent with previous standards. After reviewing the specs with company counsel, you are assured there is no legal liability, because the fine print on the order forms allows for minor variances. It is simply that the widgets won't last as long as they have in the past, which in fact may be good for business because replacement orders will begin coming in sooner than usual.

Now you can rest easy. Or can you? Something isn't right. You wonder if you should recall the widgets or issue credit memos to the purchasers. There is no legal duty to do this and it will be costly. All the while, like a Cheshire cat, the image of your lawyer keeps fading in and out of your mind's eye as he warns, "Don't open this can of worms." What do you do?

Meanwhile, on the domestic front, you are convinced the divorce case between you and your spouse makes the movie "War of the Roses" pale by comparison. Even your spouse's attorney is uncharacteristically apologetic when presenting demands. You would like to fight—every inch of the way but the contentious atmosphere is harming your children.

What is an additional five percent of marital property when compared to the emotional health of your kids? So you pay another fifty a week in alimony if that's what it takes to put an end to this state of belligerence. Your attorney chastises you for even considering such a thing. You are under no legal obligation to part with more than has already been offered. There isn't a judge in the land who would enter an order awarding what your spouse is seeking. "Don't do it!" He screams. Do you?

Situations such as these may arise where you and your attorney are no longer on the same wavelength. Or there may be other factors at work causing you to consider severing ties with your attorney. In other words, to put it bluntly, you ask yourself the question....

Do I Fire
My Lawyer?

It is not uncommon for this question to arise. But replacing one attorney with another before the matter has been concluded is like switching horses midstream. There is always a hitch.

Ask yourself why you are dissatisfied. Is it because your lawyer's advice is dubious, or you just don't like what you hear? If the latter, better think again. It may be your lawyer, trained to reason and temper the turbulence of potentially explosive situations, is doing just that, much to your chagrin. Or perhaps your attorney is serving you a dose of reality and you aren't happy about that. It is like going from doctor to doctor until you receive a diagnosis that suits you. At such times, firing your lawyer is a mistake.

On the other hand, if you have lost confidence in your attorney, it is definitely time to move on. It is one thing if your firm employs general counsel who refers and then monitors a case on your behalf, but quite another should you be tempted to follow the advice Billy Joel once offered at a workshop for musicians: "Get a lawyer. Then hire another lawyer to watch the first lawyer." If nothing else, you are paying two legal fees. Yet there are people (some astute and successful) who do this. I know because it happened to me once.

I represented a partnership comprised of five prominent businessmen concerning a condominium conversion. Anxious to make an impression and keep them as clients for the future, I quoted a contract fee about half the going rate.

After the preliminary work was completed, I circulated a draft of the condominium documents to the partners for their review. Two weeks later a set was returned with an eight-page letter from an attorney at one of Philadelphia's mega-firms. He said in the letter that he had been asked to critique the draft by one of the partners.

Seething, I dialed the phone and called the offending partner to protest. The man was shocked. He couldn't understand why I felt insulted, and explained he always retained a "reading attorney" to review the work of the primary attorney.

I do believe that if you must retain one lawyer to oversee another, you erred in hiring the first lawyer. At such times, terminate the relationship, and return to Chapter Three.

Sometimes firing an attorney involves more than a loss of confidence. For instance, take the case of TMC, Esquire. TMC was an attorney and grandson of the founder of the most prestigious law firm in his city. He was moderate in temperament, highly respected, and an old-fashioned gentleman—but a gentleman addicted to speculating in option-trading.

For a time, TMC made millions. Then he lost millions. But the millions he lost were millions more than the millions he made. To make up the difference, he appropriated the funds from his clients' estates and trusts. TMC behaved badly, which lawyers can do (Chapter One). If you suspect your lawyer is behaving badly, fire him.

The revelation of TMC's larceny came as a shock, but there are times you may have more than an inkling your attorney has an impaired conscience or is an outright crook. Perhaps you find this advantageous, say in trying to lure a competitor's key employees, or inflating a personal injury claim. Nevertheless, the rule is to fire your attorney if you know or have good reason to think he or she is dishonest or unethical.

To do this may be easier said than done.

**How To Fire
Your Lawyer**

When I owned my mortgage company, I once directed a department head to discharge an employee whose performance was unsatisfactory. It was my policy to provide severance pay rather than notice of termination, since I saw no benefit in having a disgruntled individual at work. Late that Friday afternoon, the department head and the employee met.

Monday at nine a.m., the "fired" employee entered the building, took a cup of coffee, made straight for her desk and went to work. Puzzled, I called the department head into my office. She could not understand what went wrong. She said she explained our dissatisfaction and that there was no point in continuing employment. Obviously, the message did not get through.

When discharging your attorney, be direct and precise. Do not beat around the bush. Mince no words. Do not emulate my former department head. Phrases such as "I want another lawyer," "Your services are no longer desired," or "You are fired," should suffice. If not, I guarantee that "I will pay you no more," will do the job.

Nor should you take the coward's way out and assign the task to a new lawyer. Always perform the chore yourself. Aside from being the honorable thing, it also provides an opportunity for you and your attorney to resolve what may have been a misunderstanding.

Your new attorney will generally require the existing file; therefore, there is one last piece of business to be resolved—payment of the final bill. Until this is satisfied, do not expect anything to happen with your case. Your new lawyer may be unable to proceed without the material, and in most cases would not unless assured the first attorney was paid. This practice is known as "protecting the fee" of a fellow member of the bar.

The issue of the final bill is dealt with in the next chapter, so let us assume this has been settled. Hopefully your new lawyer will prove satisfactory. If not, that means one of three things has occurred. One, you did not act in accordance with Chapter 3 in selecting your attorneys. Two, you have miserable luck. Or three, the problem is not with your lawyer but with you.

The last of the possibilities is the most likely. It is not unusual for one of the parties in a divorce case to go through three or four attorneys before relenting to the dictates of logic and compromise, which is why lawyers in this field frequently demand high retainers. However, moving from one attorney to another is not limited to domestic relations. Whenever there are unrealistic expectations, or a failure to appreciate the intricacies (and fees) of a case, finding fault with the lawyer is common.

Although firing your lawyer should never be taken lightly, if the conditions discussed here apply, do it.

Getting
Even

Sometimes discharging your lawyer is not enough. Your attorney may have really botched things up. Take the time you were on a junket in Atlantic City and were rear-ended by a stretch limousine. The owner, rumored to be Donald Trump or Merv Griffin, has liability insurance of five million dollars per collision. In truth, you were seriously hurt and in the hospital for six weeks. Your spine will never be the same. Liability is not an issue, and your lawyer assures you the case will settle in the seven figures. He is so certain of this, he has put a deposit on a beach front home in Longport.

Abruptly your lawyer does an about-face. Reaching you at home late one evening and sounding quite anxious, he urges you to settle for one hundred thousand dollars.

Naturally you are wary of this, so you contact the insurance company directly. A staff attorney informs you that the lawsuit had not been filed within the time required by the statute of limitations. The offer is the most you'll get and is only being made to avoid legal stratagems your desperate attorney is bound to take. "I strongly suggest you accept it," he closes, adding a patronizing "Have a nice day."

You don't want to fire your attorney; you want to kill him. And you want him to suffer before he dies. Although you may feel justified in executing such a deed, I do not recommend it. There are alternatives.

There was a time, not very long ago, when one lawyer suing another was unheard of. But since the 1970s, legal malpractice cases have proliferated dramatically. I know. My liability insurance premium increased 400% in a ten-year period, despite the fact I was a more experienced attorney and never had a claim made against me.

If you think the attorney you fired has made an error beyond the bounds of reasonable judgement, see another lawyer about a malpractice suit. Naturally, in selecting this attorney follow the procedure in Chapter Three, keeping in mind you want someone with malpractice experience. Therefore, this may not be the same person you hire to replace your former attorney.

In cases involving a violation of the code of ethics governing attorneys, contact your local bar association or the state disciplinary board. More and more people are now aware of this option. In just the past several years the number of complaints leveled against lawyers has increased dramatically, from 70,000 in 1985 to 93,000 in 1988.

Unfortunately, despite efforts to shore up these proceedings, the percentage of cases which result in sanctions is negligible. For example, in California where the figures reflect most states, for every one hundred complaints filed in 1989 against an attorney, only two ended in disciplinary action. And suppose the attorney is found to have misbehaved? Just what sort of punishment can you expect to be meted out?

It generally ranges from informal reprimand (naughty, naughty) to public censure (naughty, naughty and we're going to tell everyone about it!) to suspension from practice or disbarment, which does happen at times. Of course, these deliberations are under the domain of attorneys and judges (people who had once been lawyers). But that's another story.

Alas, all good things must come to a close. So it also goes with lawsuits. At some point, assuming you have not fired your attorney, you will have to extricate yourself from whatever legal entanglement enmeshes you.

CHAPTER SIX

Winding Down and Tying Up The Loose Ends

Author Thomas Pynchon issued the following admonition in his novel, *Vineland*: "According to Vato Gomez, one of the heavy-dutiest of Mexican curses goes, 'May your life be full of lawyers.'" Elaborating further, he drew an analogy between the legal system and "a swamp, where a man had to be high-flotation indeed not to be sucked down forever into its snake-infested stench." Meyer Schlemmel would have done well to have heeded the warnings of Pynchon and Vato Gomez.

Meyer Schlemmel was a balding, middle-aged man who received more in life than he had ever asked or expected. His attractive wife and two teen-aged children lived with him in a spacious suburban home. He was the owner of a thriving company which sold hardware to the government, a minuscule cog in the vast wasteland of the military-industrial complex.

Schlemmel's troubles began when a routine internal audit disclosed one of his bookkeepers had embezzled a few thousand dollars. Naturally, the culprit had to be fired, but Schlemmel thought it best to consult his attorney so that the matter would be handled tactfully and properly. His lawyer urged that in order to maintain the company's integrity, not only should the employee be immediately discharged, but litigation must be commenced to seek restitution.

"But wouldn't it be better to keep this quiet?" Schlemmel asked. "Anyway, I don't want to ruin this woman's life. It was a mistake. Losing her job is punishment enough." Nonetheless, the attorney prevailed and the suit was filed. One day, several weeks later, Schlemmel's regular morning respite of bagel and coffee was rudely interrupted by a commotion outside his office. Opening the door, Schlemmel was confronted by a boisterous man arguing with his secretary that he must see Mr. Schlemmel personally to deliver an urgent package. "Well, young man, now that you found me, what do you want?" Schlemmel asked.

As it turned out, the man was a process server, and the urgent package a set of legal papers from the bookkeeper's attorney. With heartburn searing his insides, Schlemmel faxed the material to his lawyer, who informed him that it was merely a counterclaim for wrongful discharge of employment and discrimination.

"Discrimination!" shouted Schlemmel. The bookkeeper, who was Hispanic, alleged that to be the real reason for losing her job. "Maybe we should just forget the whole thing," Schlemmel said, but his attorney insisted they stand firm.

The case began to receive some notoriety in the local press. Schlemmel was portrayed as a ruthless bigot who picked on an innocent employee to cover up his own embezzlement. Meantime, unbeknownst to Schlemmel, a covert section of the federal bureaucracy assigned to reading the nation's newspapers in the hope of turning up something on someone at sometime violating some sort or any sort of federal statute, regulation, or whim, caught note of Schlemmel's problems and ordered an audit. As a result, one gray morning two auditors from the U.S. Inspector General's office appeared in Schlemmel's office and gave the company a thorough going-over.

Three grueling weeks later, everything was found to be satisfactory, with one exception. The Hispanic bookkeeper was an illegal alien and Schlemmel as her employer was debarred from further dealings with the government for six months.

"That puts me out of business!" Schlemmel screamed at his lawyer. "Then we'll sue the government," the lawyer said, "But it will take some time." "Why?" Schlemmel asked. "Because we need their permission first."

The tabloids had a field day with the disclosure that the bookkeeper was an illegal alien. Schlemmel was accused of being an exploiter of wetbacks who then preyed and profited on the taxpayer. Mrs. Schlemmel was snubbed at the club and the Schlemmel children were taunted at school. All was not well in the Schlemmel household.

Anticipating a serious cash-flow problem, Schlemmel met with his bankers to increase his line of credit. Given the lawsuits and publicity, not to mention a standstill in operations, he was turned down. Dejected, Schlemmel conferred with his attorney and once more he pleaded to put an end to all the suits and countersuits. Instead, she referred Schlemmel to another attorney, one specializing in bankruptcy, to file Chapter Eleven and buy some time.

Wearily, Schlemmel returned home. He had laid off half his staff, approved the filing of a bankruptcy petition, testified at a deposition on the case with the bookkeeper, and hand delivered tons of files to both his attorneys. All in one day. He was dead tired. Thank God to be home.

Schlemmel opened the door to his haven from the world's craziness and as was his custom for two decades, he cried out, "Hi hon, I'm home." No answer. As his gaze scanned the dwelling there was nothing to see but barren walls and empty rooms. No wife, no kids, no furniture, no nothing. Just a note taped to the banister explaining he would hear from his wife's divorce attorney.

"Stop it! Bring these damn cases to a close! I don't care what you have to do—do it!" Schlemmel shrieked at his lawyer. In the end, all he got was the name of a "good" divorce lawyer.

After a time, lawsuits have a way of assuming a life of their own. They and not you take control, and like a tornado they can consume everything in their path—even you, the client. This is what happened to Meyer Schlemmel. Don't let it happen to you.

There is nothing you can do to change the complexities and intricacies of the law. What appears as a simple matter often is not, and it will take time to arrive at a conclusion. However, like a good novel, there is always a climax, and once this occurs you must keep the momentum going.

Lawyers often assume more work than they can handle. As a result, the more pressing matters are attended to while the rest are put aside. So it will go with your case once it has reached a resolution or its "climax."

This is when you must apply all you have learned. You are the Boss (Chapter Five)! Remind your attorney, who is no more than a mere mortal (Chapter One), of the Ground Rules (Chapter Four) and that you want the case expeditiously laid to rest. If necessary, assert yourself. Do not be a "schlemmel."

It should be so easy. If it were, I would not have fallen victim to this very pitfall not so long ago.

The circumstances involve the case of a residential housing developer and Township X. In addition to serving as general counsel, I was an officer and shareholder of the development company. A lawsuit was brought against the township after the project was completed. Because I no longer engaged in litigation, I referred the matter and we retained a firm known to be aggressive when representing builders against municipalities. Soon, a fifty page multi-count complaint was filed against Township X, its commissioners, and the township manager.

After firing several legal salvos over the course of a year, an accord was reached, which was reduced to a court-approved stipulation. Several matters remained to be worked out but our attorney convinced me to allow the township time, and if they did not comply with the stipulation, we could return to the judge and obtain a contempt order. I consented. Wrong.

It's like going to a dentist to have a cavity filled. There you sit squirming, getting your gum stuck with a needle full of novocaine while your gaping mouth is being packed with cotton rolls, metallic devices and clamps, and all the time the dentist is cracking jokes to the hygienist. Finally, the drilling begins and a smell like burning flesh assaults your nostrils. Just before you are convinced you will salivate to death, the dentist discovers another cavity in an adjacent and already frozen tooth. "Might as well get it over with," he shrugs. "After all, why bother coming back again?" Your grunt is taken for acquiescence.

Same with a lawsuit. When you are in court or at the bargaining table, get it over and done with. Don't allow yourself, as I did, to be put off. In other words, don't be "schlemmeled"!

As I suspected, Township X dragged its feet and the case lingered on. The township solicitor and our attorney had more urgent business. In the same way my lawyer dropped everything to answer my calls when our case was in the critical stage, now

I was the one being moved to the back burner, waiting three and four days for a response to a phone call and weeks to a letter.

Of course the law firm continued to submit monthly invoices, including such items as file review, status update memo, inter-office conference, and "returning" my phone calls. These bills managed to generate three, four, or five hundred dollars per month for essentially accomplishing nothing. Finally, I took charge, in the hope of promptly resolving the matter.

The next month our lawyer's bill consisted of time spent in reading copies of my letters to the township solicitor. I stopped furnishing him with copies. Not hearing from me, he sent me a letter requesting he be kept abreast of developments. This letter cost sixty dollars. And so it went. Things were getting desperate.

Only after I set a deadline to return to court and seek a contempt order did Township X cooperate. An additional year and several thousand dollars in legal fees had been spent to conclude the matter, both of which could have been avoided if there were no loose ends left when the stipulation was made. So, tie up those loose ends!

"A Lawyer's Time and Advice Are His Stock and Trade"

Thus counseled Abraham Lincoln, still another attorney who became president. Not only was the gangly, bearded statesman a politician, president, and rail-splitter, but he was also a lawyer—and a darn good one at that. For the one million lawyers practicing today, his credo probably adorns the walls of more law offices and has been instilled in the hearts and minds of more lawyers than all other legal adages combined. As a result, attorneys are free with their time though their time is not free for you. Do not become trapped in a lawyer's time warp. It is very costly.

You probably cannot fathom how a fairly intelligent person can be beguiled by an attorney into drawing out a lawsuit for the purpose of generating additional fees. Yet far more savvy and sophisticated clients than the average consumer or small business enterprise have been hoodwinked by lawyers out of thousands or, in the case of Attorney LBS, millions of dollars.

Attorney LBS practiced law in California. He had been called, by other lawyers no less, the most litigious man on earth. In one case, he sued a neighbor for a quarter of a million dollars over a parking space in front of his house. However, a zeal for suing is not illegal nor is tardiness in bringing a case to a conclusion. But a conspiracy to do this for the purpose of billing clients is a crime, as LBS discovered when he and eighteen other defendants, fourteen of whom were lawyers, were indicted by a federal grand jury for defrauding their clients of $50 million by manipulating and churning litigation.

Who were the naïve clients who fell prey to this scheme? None other than Fireman's Fund Insurance Company, Allstate Insurance, and thirty-one other insurance carriers, who shelled out millions after millions in legal fees because of a variety of stratagems designed to "complicate, prolong and raise the cost of litigation." Some of the methods employed by these unscrupulous lawyers included charging two dollars a page for photocopying, billing $120 an hour for a lawyer who was contracted by the law firm at an hourly rate of $20, and posting time-slips for twenty-four-hour days.

The point is that it can be difficult to get your lawyer to put an end to the case and bid your checkbook *adieu*. If you find yourself mired in such a situation, do not feel helpless nor alone. It happened to me, and to Allstate, and to Fireman's Fund, and to many, many others. Don't let it happen to you.

The Final Bill, or
Client in Wonderland

At last it has arrived! Your lawyer's last bill. Perusing the statement and trying to make sense out of what appears to be written in Jabberwocky leaves you feeling like Alice peering through the looking glass. If this describes you, take a cue from Alice who was never shy about asking questions. Stand up to your lawyer just as Alice confronted the red Queen. The risk of losing your head is not nearly so great, as the likelihood you will part with money which rightfully should remain yours.

After carefully examining the invoice, there is something you must do before contacting your attorney. You must prepare. If you do not, you have no chance at all against one who has grown accustomed to justifying bills and does so in a language combining legalese and gibberish.

In order to be ready for this encounter, review either the engagement letter or the contingent fee agreement, one of which you should have received if you followed my instructions in Chapter Four. In addition, refresh your recollection of the ground rules and any conversations you and your attorney had concerning the fee.

Now study the bill again. It should conform to what you and your lawyer had agreed upon. If it is a contract fee, make certain this is all you are paying. In the event you consented to a "value" consideration, get ready for some hard haggling and a good amount of jawboning. A fee based on a percentage is a simple question of mathematics. If a contingency, make sure it corresponds to your prior understanding and is in accord with the contingent fee agreement.

When an hourly rate is involved, nothing less than an exhaustive examination of the bill will suffice. Do this even if your lawyer is returning a portion of an "unused" retainer (perhaps more should have been sent back). And, in the event you are being asked for more money, do this twice. Be on the lookout for any discrepancy between your records of conferences and phone calls with your lawyer and what is billed. Any significant variance suggests that all is not well. Although you cannot count on red flags, here are some line items which should make you suspect:

Review of file; review of letter; review of country club membership list (attorney's explanation—to see if there is an "in" with the judge); inter-office memo; inter-office conference; research (find out the purpose and ask for a summary); phone calls to judge's law clerk (what judge? this was a real estate closing!); conference with your spouse's counsel; conference with your spouse (?!); preparation of bill....

"I'm being charged for preparation of the bill!" Yes, it does add insult to injury but this is a common practice among many attorneys. Frankly, I think this is unconscionable and should never be paid. Its rationale brings us back to Lincoln's maxim for lawyers. Any time spent on your file is billable, even preparation of the bill. Somehow, I don't think this is what Honest Abe had in mind. If push comes to shove on this issue, shove.

How to shove brings us to the next portion of this chapter. What do you do when you and your attorney are at odds over the bill? You have several choices.

Confrontation
and Compromise

With your courage stoked by the knowledge that "justice" is on your side, you pick up the phone and call your lawyer. In a tone marked with civility and candor, somewhat removed from the previous milieu of camaraderie, you tell your lawyer you don't agree with the bill.

The reaction you might expect will vary depending upon the personality of your attorney and whether the bill is accurate. So, you may hear any one of the following responses:

"Look, calm down and then come on in so we can discuss this."

"Harrumph, harrumph... Uh hem... can I uh get back to you?"

"I'm sorry you feel that way but there is nothing further to talk about. The statement is final. I *expect* to be paid." (And, if you don't pay, you can expect to be sued.)

"You don't want to pay it? Then don't! Forget it! See what I care. And after all I did for you.... "

"Tell you what, just send me what you feel is fair. If it seems right to me, that will be the end of it."

"I don't have the file handy just now but I'll review it as soon as I can. Then I'll send you a revised bill or an explanation of the charges."

"Really? No shit ... I'll have to go over this with bookkeeping. Maybe they fouled things up. We're on computers and things like this can happen ... but I'll check it out. Hey, how ya been?"

"Oh... yeah... sure ... uh listen... uh just take off 25% and send me the rest ... ok?"

"Same to you fella!" (hangs up).

If you were profane or abusive at your end of the conversation, you can expect to receive the last retort, which you would justly deserve. However, unless the attorney agrees with you, the question becomes what to do next. My answer is unequivocal—negotiate and compromise.

Remember the lawyer in Chapter Five who almost killed the deal for the client, and where I was asked to resolve the matter of his outstanding invoice? In fact, I only played an advisory role and it was the client who conducted the actual negotiations.

The problem arose because of failure to have a clear understanding between attorney and client concerning remuneration. The attorney originally quoted a contract fee of $8,000. When the client balked, the attorney agreed to revise it downward, depending upon how much time would be spent, and assuring the client with a "trust-me" smile that there would be no problem. I suspect this is what prompted the lawyer to make things more complicated than they were. In any event, after he was dismissed, a final bill was submitted for $2500.

At my suggestion, the client began negotiations by politely requesting an itemized statement including specific services rendered, time spent on each, and the hourly rate. This was not furnished. Instead, in a sanctimonious letter, the lawyer wrote that the amount was reasonable and payment should be made. The attorney had played his hand. He could not support the amount he billed with anything other than his opinion that it was "fair."

I knew this for what it was. Every first-year law student is taught the same strategy. When in court, argue the law. Should that not succeed, then argue the facts. If that fails, pound your fist on the table and scream for justice.

And that was precisely what this lawyer was doing. I advised the client that in my opinion no more than ten to fifteen hours could have been spent on the matter. An hourly rate of $150 was probably what the attorney charged.

Therefore, a check in the amount of $1500 with an endorsement on the back indicating "settlement in full" and a cover letter should end the matter. It did. $8000 became $2500 which in turn became $1500. Negotiate and compromise whenever you can.

Sometimes differences over the fee cannot be settled amicably. Then what?

The Fee-Dispute Committee,
or "Justice Lawyer Style"

Picture this. You enter a small oval conference room situated in the local bar-association headquarters, an edifice dating back to 1796. The building was first used as a tavern and inn. Subsequently, it served as a local meeting house, then as county seat, later as the clandestine command post of the Ku Klux Klan, once again the county seat during construction of a grander structure, and from 1929 to the present, home to the county's nine hundred and seventy-four lawyers. A building which has seen drunkards, gamblers, politicians, and racist murderers as its previous occupants. A fitting site for the local barristers if there ever was one.

Back to the oval conference room. You are asked to take a seat at a round table. Your former attorney, Clyde "Buddy" Goodfellow, is seated to your left. Filling out the table's circumference are three members of the bar association's fee-dispute committee.

First there is Kenton D. Fitzworth III, Esquire, whom you guess to be forty going on seventy: A starched shirt. A three-piece blue pin stripe suit. A foot shod in a wing tip Cordovan is impatiently tapping the floor. Eyes glare at you, suggesting he has better things to do. Kenton D.Fitzworth III, Esquire, is definitely not on your side, you conclude. Two votes left.

Mary Bailey extends her hand and warmly smiles as she introduces herself. She is black and young. Bound to be sympathetic and understanding. Certainly sensitive to those who are taken advantage of by others in a more powerful position. This one is in your pocket, you convince yourself. That is until she shakes Clyde's hand, laughs, and turning aside but not far enough to prevent you from seeing, winks. Your heart sinks.

Barely awake is Emmanuel Schwartz, his bald head slumped forward. From an open collar a bright red tie hangs loose and limp, partially covering the gaps in the shirt where two buttons have popped. Mary introduces Emmanuel to you as the senior member of the committee. If there is to be a swing vote, which appears more and more unlikely, Manny will be the one to cast the deciding ballot. Wonderful.

The way you figure it, your situation is strikingly similar to what may have happened a hundred years ago to one of Mary Bailey's ancestors, who was wrongfully accused of stealing a mule. Stealthily and in the dead of night, Great-granddaddy Bailey was snatched from his shanty by a mob of white-hooded hoodlums and then carried to this very building. You can see it all now. The hangman is busy securing the knots on the rope, every so often giving the noose a whirl in the air. You are convinced that your case is lost before it began.

Despite an inclination to dismiss fee-dispute committees, almost always composed exclusively of lawyers, as inherently prejudicial, you will find that there is some merit to this option.

Most often, as with arbitration panels, the decision reflects a compromise. However, while there are times when the attorney is awarded the entire amount, the reverse does not hold true, and you should not expect to walk away without having to pay at least a portion of the fee.

These proceedings are conducted informally and legal representation is not required, although if the amount in controversy is substantial, it may be wise to consult an attorney for guidance or have one attend the hearing with you. What matters most is that you be prepared with documentation to support your position. Complaining that your lawyer did a lousy job is pointless, and the panel will hear none of it. If this were a legitimate reason to deny compensation to an attorney, half the lawyers would never earn anything.

You should certainly consider using a fee-dispute committee to resolve any differences over the last bill received from your lawyer. Check with your local or state bar association to determine if it is one of the majority of states which have a mechanism like this in place, and if so find out the procedure involved. It is a simple, relatively swift, and inexpensive route to take. However, there is one significant drawback. You may have to sign a legally binding document that you will accept and abide by the judgement of the tribunal. In other words, this is the court of first and last resort. If you are unhappy with the results—tough. There is nothing further you can do about it. The fee-dispute committee's decision will be upheld and enforced in a court of law. If they say you must pay—you pay.

So, what do you do? Should you turn to the services of a fee-dispute committee or not? I suggest that if it is non-binding, then by all means use it. If not, you should still give it consideration, and possibly consult another lawyer before reaching a decision. However, if your former attorney is one of the "good old guys," a member of the bar for many years, and probably known by the panel of attorneys who will hear your case because it is a small bar association, think again.

I did.

Do Unto Thy Lawyer As Thy
Lawyer Would Do Unto You—Sue!

Archibald MacBribe, Esquire, was a well connected attorney in a sprawling suburban county ripe and poised for real estate development in the early 1970s. Over the three decades he practiced, Archie (as he was called by the local politicos) developed an expertise in zoning law and held the post of solicitor to several township zoning boards. The word was that if you wanted a lawyer to represent you concerning real estate development, Archibald MacBribe could get it done in the wink of an eye.

And therein lies the giveaway—the eyes of Archibald MacBribe. No matter how long you and Archie were engaged in a conversation, not for an instant did he ever look you in the eye. Not me, nor my clients whom I had referred to him, nor opposing counsel, nor anyone in the world. Archibald MacBribe never looked anyone in the eye.

Although I haven't mentioned it until now, "Absolute and Unconditional Rule Number One in Dealing with Lawyers and All Other Sorts of Human Beings" is this: Never trust anyone who doesn't look you in the eye. This I learned too late, after Archibald MacBribe, Esquire, tried to cheat my client out of $5000.

Since I had been practicing law for less than two years, most of it spent in the district attorney's office, I was hesitant to assume the responsibility of representing this particular client in obtaining a zoning change on two hundred acres from rural/farm to multi-family residential. MacBribe was hired to obtain this result. His retainer was $5000. He met with the client for about an hour, attended one township planning commission meeting, and then was dismissed for his conduct arising out of the following situation.

On another matter with the same client but in a different township, I did feel competent to take on what was anticipated to be a rather minor zoning change from half acre to quarter acre lots. The township officials were receptive and cooperative. However, I began to suspect a problem when on the night of the hearing I could not secure a parking space in the municipal building's lot. Inside, a hundred boisterous neighbors waving homemade posters and signs resoundingly made their opposition known. And in the midst of the multitude was their attorney, none other than Archibald MacBribe, Esquire.

This is the most blatant case of a conflict of interest I ever encountered. MacBribe tried to justify it by claiming he was a resident of the community and only exercising his rights as a citizen. No matter. On behalf of my client, I asked for a return of the unused portion of the retainer.

Archie refused.

In weighing the alternatives, including the county bar association's fee-dispute committee, I sought out the opinion of my law partners. Although we were young, inexperienced, and naive, we were not stupid. Archie was definitely one of the "good old guys" (at the time, there were no "good old gals") and a senior partner at one of the most prestigious firms in the county. Since the decision of the fee-dispute committee was conclusive, we advised the client to sue MacBribe for a return of the retainer in its entirety, given the conflict of interest.

The complaint was filed and the pending trial became the gossip of the local legal community. My partners and I were depicted as brash and abrasive city-bred lawyers (though we all lived in the suburbs) who had the audacity to sue another attorney. Archibald's indignation and dander was up. There was no movement to compromise and settle. The trial was at hand.

Since I might be called as a witness, one of my partners tried the case. Everything proceeded in a composed manner until Archibald took the stand in his defense. Fumbling with hastily scrawled time slips, Archie's eyes betrayed his lips. They were at their best as they drifted in all directions. When questioning him, his lawyer, a partner in the same firm, skipped about the court room trying to at least give the appearance that Archibald was looking at him.

Archibald MacBribe was lying. His tally of time per hourly rate, which coincidentally totalled $4,955.00, did not impress the judge. The judge didn't believe a word Archie was saying. Archibald MacBribe lost the case. He was ordered to return all but $900 which he was allowed to keep only because the judge gave him the benefit of the doubt on the conflict of interest issue.

When the amount in contention is very substantial, going to court may be the only alternative. This was the case involving the law firm of EF&M suing its client for almost $1 million in unpaid legal fees. If the mere fact that lawyers would allow a client to owe such an amount makes you suspect, you would not be alone. No doubt this was one of the reasons leading the judge and jury to conclude that "the bill was grossly inflated and without justification." EF&M was awarded only forty percent of the amount at stake.

So you see, justice can be found in our courts of law even if the opposing party is an attorney.

Of this you can be certain. If you wind up in court with your former attorney, it is axiomatic that said attorney will never work for you again. Lawyers are people (Chapter One) and like all people, lawyers hold grudges. Also, like all people, lawyers do not relish being sued. So, suing your lawyer or forcing your lawyer to sue you for a fee will leave you with one less question in life to answer: "Do I go back to the same lawyer?" This arises, along with other considerations, as you proceed to the next case.

CHAPTER SEVEN

The Next Case

During the mid and late 1950s, Hollywood inundated the American moviegoer with a deluge of black and white "B" horror movies. These films were variants on the same theme: Mankind (women were absolved), in its folly to pursue the use and deployment of atomic weaponry, made an "oops."(An "oops" is defined as what little kids do when they drop things, or grown-ups mutter when they say the wrong thing, or doctors breathe under their face masks when the scalpel slips.)

In any event, this genre of films always had an "oops" occur somewhere near the beginning, which resulted in the spread of radioactivity all over something or someone. The result was a hideous, generally homicidal mutant or race of mutants.

Though this took many forms, the favorite seemed to involve the transformation of insects from tiny innocuous creatures into huge menacing anthropoids with an insatiable appetite for human flesh. For me, the most hideous of this collection, the most fearful and awesome, was the transfigured predaceous arachnid, or Giant Tarantula.

The way I see it, this is how it is with our litigious society— controlled and exploited by lawyers, a cross between the cunning and crafty spider and the towering tarantula. Therefore, although you may succeed in extricating yourself for a time from the well woven web of the law, don't expect it to last.

In other words, the answer to the question, "Will I ever need a lawyer again?" is

You Can
Count On It!

Relying upon what you learned in Chapter Two, first determine if you need a lawyer or whether you can go it alone. Once you conclude legal representation is appropriate, you must then decide whom to hire.

Review Chapter Three. Abide by the Four Thou Shalt Nots:
• Thou Shalt Not yield to the temptation to ask for free representation.
• Thou Shalt Not permit thy fingers to do the walking through the Yellow Pages.
• Thou Shalt Not covet a law firm because of its largeness.
• Thou Shalt Not seek a politically connected attorney.

To locate an attorney, you may wish to consult the *Martindale-Hubbell Law Directory* or the bar association lawyer referral service. Pay attention to reputation and word of mouth. Weigh the advice of others such as business associates, your accountant, your banker, friends and family, and even other lawyers.

Finally, when the list is down to two or three candidates, meet them. Discuss what the cost will be, and don't hesitate to say no if you can't afford that particular attorney. Try someone else.

Depending upon the complexity of your case, you may want a lawyer who specializes in that type of work. Of course none of this will mean a thing if the lawyer does not have the time to perform the job adequately or lacks the staff and personnel to represent you properly.

While it appears I am telling you to return to Chapter Three, this is not so. There is a new ingredient which has been added to the recipe for selecting the right attorney which should be obvious. What about the lawyer who represented you on your previous case?

Do I Go Back to
the Same Lawyer?

During the summers, I love to walk along the beach to the point where the ocean meets the bay. There the raucous ocean current is stilled by a natural stone jetty extending several hundred feet, and separating the two bodies of water. About halfway out, an old man sits on a striped canvass chair, his line stretched into the sea and a rusty can of minnows precariously balanced on the rocks by his side.

As I approach, I wave and shout, "Catch anything?" The old man raises his head and flashes a toothless grin.

"Nup," he always answers.

This has gone on for years, and although I am not a fisherman, it doesn't take a genius to figure that maybe he should consider a new location. So one day last summer instead of just smiling back at his customary "nup," I ventured out on the rocky embankment to speak with him.

"Excuse me, but there is something I've been wanting to ask you." The old man gazed at me. "Why do you always fish in the same spot when you never seem to catch anything? Don't you care? Or do you just come out here for the sea and salty air?" The old man's wizened bronze skin crinkled into a puzzled expression.

"Sure I want to catch fish!" He cackled regarding me as if I was crazy.

"Then why keep coming back to this spot?"

"Don't know," he said, removing his captain's cap and scratching a frazzled head of white hair. After a moment's pause of staring up into the sky, he went on. "Guess it's cause I been comin here for thirty years. First time out caught a flounder, big one too. Not much since though." He hesitated, serenely peering around and then added, "Just keep comin out of habit, I suppose."

Should you go back to the same lawyer? Maybe. But never do it out of habit. The problems in changing attorneys before your matter has been resolved which were discussed in Chapter Five are not encountered here. The issue now is whether you were pleased with the services previously rendered by this lawyer.

To answer this may be more difficult than it seems. Therefore, I have devised a test to evaluate your former attorney.

How To Score
Your Lawyer

Instructions: Answer every question or statement with a simple "yes" or "no." If in doubt, answer with the one which most applies. If still in doubt, leave it blank, but this will go against the attorney. Knowing this, if you do leave a question unanswered, the lawyer deserves a negative score.

The test is divided into categories. Use a number two soft lead pencil with a fresh eraser. Make certain the pencil point is sharpened, and if it breaks deduct two points from the final score. That would indicate merely thinking about your lawyer makes you mad or upset.

There is no time limit, so work as leisurely or as quickly as you desire. If it were me, I'd get it over with in a hurry. Who wants to spend more time than is necessary thinking about lawyers?

Ok. Here goes and good luck (for your former attorney!).

The Comfort Quotient

1. Did you feel harried or under pressure whenever speaking with your attorney?
2. Were you and your attorney on a first-name basis before the conclusion of your case?
3. Did you break into a sweat whenever you met, spoke, or even just thought about your lawyer?
4. Does your attorney have a fine set of teeth? (If you can't answer, it is because your attorney never smiled. This is one of those tricky questions, and a blank answer counts against the lawyer.)
5. Did you confide anything to your attorney you never told anyone else?

The Competency Quotient

6. Did your attorney keep asking you about things which you previously explained? (If your attorney was elderly, it may be due to senility, which is not recognized as incompetency under the law. If it were, several members of the United States Supreme Court would be declared incompetent.)
7. Was your lawyer ever berated by the judge in your case? Or if there was no trial, did opposing counsel or other parties involved have a tendency to roll their eyes every time your attorney said something?

8. Did your attorney tend to stammer, stutter, or speak in an otherwise unintelligible manner? (Answer no if due to a speech impediment.)

9. Did your attorney often request continuances for hearings or reschedule meetings?

10. Did you discover mistakes in letters or documents prepared by your lawyer, other than typagrafical (sic) errors?

11. All in all, do you feel your attorney knew what he or she was doing?

The Accessibility Quotient

12. When calling your attorney, you were always screened by two or more secretaries before being put through.

13. While waiting for your lawyer to pick up the phone (answer "yes" if any one of the following applies):

(a) you heard one entire song on the tape.

(b) the prerecorded message reciting how very important you are but please be patient and Attorney Soandso will be with you in a moment kept repeating itself.

(c) the sonorous voice on the tape lulling you into a trancelike state was beginning to sound like someone you knew.

(d) by the time your attorney answered, you forgot why you called.

14. Your lawyer always returned your calls within twenty-four hours.

15. When scheduling an appointment with your attorney, you heard phrases like, "Let me see when I can squeeze you in..." "How about 3:15 to 3:25?," "I'll put you through to my secretary to schedule a time," "Call me Tuesday and we'll set something up."

16. Your attorney never canceled an appointment.

17. All in all, you had the impression your lawyer wanted to see you about as much as a doctor looks forward to making a house call.

The Success
Quotient

18. The last time you saw your attorney, you smiled and warmly shook hands.
19. In explaining the outcome and seeming a bit on guard, your lawyer proclaimed in a patronizing tone one or more of the following: "The Law isn't perfect" "Everything considered, we got the best we could," "What did you expect?" "I never promised you that!" "Sometimes you just have to make do ..." "What goes around comes around" "Things always work out for the best" "It could have been worse" "Don't worry, we'll appeal!"
20. By the time your case came to an end, you developed one or more of these conditions: a nervous tic; stomach pains; insomnia; a general feeling of malaise and lethargy causing you to stay in bed past noon; sudden impulses to do harmful things to yourself; an increasing perception the world is out to get you (paranoia); a recurring nightmare that you have been convicted of murdering your attorney, and as you wait to be escorted to the electric chair by a Catholic priest (you're presbyterian) bearing an uncanny resemblance to the late actor Pat O'Brien, the apparition of your lawyer's ghost appears to inform you the governor has rejected your request for a stay of execution.
21. All in all, you were better off at the conclusion of your case than when it began.

The Value
Quotient

22. Your attorney's total fee was consistent with that set forth in the engagement letter. (If there was no engagement letter, answer "no".)
23. When paying your attorney's bills, you exhibited one of the following: forgetfulness, such as not placing a stamp on the envelope or neglecting to sign your name on the check; an inability to perform basic mathematical calculations; psychosomatic manifestations such as headaches, nausea,

dizziness, and the like; a sense of wonderment and awe that someone can charge so much for so little so many times and get away with it so often.

24. The money your attorney obtained for you or helped you avoid paying was more than twice the amount of the total fees and costs you incurred (Answer yes if not applicable. We'll give the lawyer a break on this one).

25. All in all, though not as gratifying as slipping a lemon past a car dealer on a trade-in, you feel you got your money's worth with your lawyer.

Now Tally
the Points

Where you answered with a "no" for any of the following questions or statements, add a point for your lawyer: 1,3,4,6,7,8,9,10,12,13,15,17,19,20,23. On the other hand, give your lawyer a point where the answer was "yes" to the following: 2,4(*),5,11,14,16,18,21,22,24,25. (*the lawyer gets a point here either way, unless the answer was blank)

Next step. Add the number of points your attorney earned. Double check. Remember, if you broke the tip of your pencil, subtract two points from the score. If the total number of points appears to be higher than what you think your lawyer deserves, deduct anything between one and three points. This reflects your "gut feeling," which is probably the most important factor. Conversely, should the score seem a tad too low, add one point but no more. Lawyers are never entitled to the benefit of much doubt. One point is plenty.

Evaluating
the Results

Excluding the "gut feeling" adjustment, the maximum number of points an attorney can earn is twenty-five. Based on the score received by your former lawyer, refer to the appropriate category below.

20 points and above. If your lawyer landed in this group, what is there left to say other than "a job well done." There is no reason you should hesitate going back to this lawyer. Even if your case involves a specialty which the attorney does not possess, the chances are an attorney with this high a rating will be the first to refer you to someone experienced in the appropriate field.

Several attorneys who placed in this category on previous test results are: Daniel Webster, Clarence Darrow, Perry Mason, Alan Dershowitz and Louis Nizer.

15–19 points. This class of attorneys merit your serious consideration when you are once again in need of legal representation. However, unless your lawyer scored at the upper end of this bracket, there is no presumption in his or her favor. Instead, be guided by the criteria set forth in Chapter Three regarding the type of lawyer you should select.

If you are still on the fence, it would be wise to analyze the test results at the "Quotient" level. For example, let us suppose the attorney performed particularly well in the Competency, Success and Value Quotients but below par in the Comfort Quotient. If you do not mind having a poor personal relationship with your lawyer so long as the job gets done, then stay with that attorney. On the other hand, Comfort and Competency Quotients high but Value and Success scores low would dictate seeking another attorney, unless you are indifferent to results.

You should fare well if you base your decision upon this test and Chapter Three. Some of the lawyers who have made the 15–19 grade are: Abraham Lincoln, F. Lee Bailey, William Kunstler, the entire firm of L.A. Law, and my brother-in-law, who is a litigation lawyer in Philadelphia.

10–14 Points. There is an adage to which I have always adhered. It goes like this: "If you do me wrong once, shame on you. But if you do me wrong twice, shame on me." So it goes with the lawyers whose scores placed them here.

If you return to an attorney who belongs to this group and you are unhappy with the services rendered the second time around, you deserve what you received. You never should have gone back. For shame.

This is not to say that circumstances might not exist where you would rehire an attorney of this caliber. What can be said, however, is that the past experience was neutral, if not downright negative, and there should be very compelling reasons to offset this fact.

If you do decide to consider this lawyer, scrutinize the test results section by section. Perhaps your attorney earned the maximum points for Success and Competency but was otherwise a high-priced SOB, yet you care only about the bottom line. Or it could be that the new matter for which you require representation suits this attorney more than the former case.

For example, the attorney in question previously represented you in delicate contract negotiations and, behaving like a bull in a china shop, really botched things up. But this time you need a lawyer to collect overdue accounts from customers you never want to do business with again. You don't care how it gets done so long as the money starts coming in. In such a case, a bullheaded, surly lawyer fits the bill.

Despite new circumstances which may justify returning to an attorney in this category, if you hire this lawyer again, do it with your eyes wide open.

Some attorneys who occupy the ignominious 10–14 stratum are Aaron Burr, Huey Long, or any one of Al Capone's many legal mouthpieces.

Less than 10 points? Hardly possible. If you did not fire this attorney the first time around, you have already made a major mistake. The fact you are even considering returning to one of the dregs of the legal profession reveals a significant personality or psychological disorder (with you, not the lawyer). Perhaps you are a masochist? Or you were fixated at the anal level of development and simply cannot let go of anything, not even such a louse as this. Or possibly your perception of the world is not in tune with reality and it took this test to reveal your attorney's true nature. Whatever. See a shrink if need be, but under no circumstances should you go back to this feeble excuse for an attorney.

A sample of the sort of creatures comprising this pack of legal chameleons are Judge Roy Bean, who at one time was the only law west of the Pecos, dispensing justice with a bottle of booze, a shotgun, and a rope; and the infamous Friedrich von Burstenburg IV, the Prussian barrister holding the world record for representing the most Nazis at the Nuremberg War Trials and who departed from the standard defense that the accused were only following orders and instead urged the tribunal to acquit his clients because their acts were justified given the natural laws of Aryan racial superiority.

But Suppose You Just Don't
"Feel" Like Going Back?

While I'm not big on sticky confections, there is one on the market that does fascinate me. It is a chewy candy shaped like a shark, with a label reading: "Gummy lawyers: Like the real thing, they'll leave a bad taste in your mouth." And so it might be with your attorney. Yet perhaps it's not the lawyer's fault that the memory you have is so negative it brings about a burning sensation in your chest, bearing all the symptoms of heartburn.

Let me put it another way. As a young boy, I once got sick after eating two or three lox and bagel sandwiches. Despite the fact I knew my nausea was caused by a virus, lox and bagel became the culprit. Guilt by association. I simply couldn't eat it.

I recount this episode from my life because it parallels what frequently takes place between attorneys and their clients. This is not something I had the insight to detect on my own. Rather, it was suggested to me by another lawyer, Attorney Doe, with the following incident.

Attorney Doe represented John Smith as both his personal lawyer and general counsel to the thriving manufacturing firm which had been in Smith's family for generations and which Smith now headed. As counsel to the company, Doe was efficient and effective, whether drafting contracts, collecting receivables, advising on employee relations and benefits, or negotiating terms with banks.

In Mr. Smith's private life, Attorney Doe was confidant and friend. He prepared estate plans; revised wills; established trusts for the children; and handled minor altercations with contractors who furnished services at the Smith estate. In other words, Attorney Doe was someone John Smith could count on. And he did.

But never with the same sense of urgency as on the day he mysteriously summoned Attorney Doe to his office.

"I need to see you, that's all. And hurry," Smith said in an agitated voice. Dutifully, Doe arrived within the hour to learn from the chagrined CEO that his secretary (and mistress) of two years, was now putting the screws to her boss. With Smith unwilling to divorce his wife in order to marry the secretary, she was threatening to write letters to Mrs. Smith revealing the adulterous affair, and to the IRS disclosing questionable tax write-offs, including trips out of town when she accompanied her employer.

"You have to straighten this mess out for me, Doe. Do whatever it takes but just take care of it. Please ... "A rueful Smith pleaded of Doe. And, as usual, Doe did.

To forestall any further complications in the matter, Attorney Doe made suitable arrangements with the secretary to begin work in a better-paying position with one of his other clients. Covering all the bases, Doe then prepared an exit memorandum providing sound, substantial reasons for her dismissal, which was signed and placed in the file.

Attorney Doe had done his work well. Smith was off the hook. Although he did not expect additional financial remuneration above his hourly rate, he did anticipate a demonstration of appreciation on a more personal level. It never came. In fact, much to Attorney Doe's bewilderment, his good friend and client appeared to be avoiding him, even going so far as not returning the vexed attorney's phone calls. For all intents and purposes, Mr. Smith seemed to drop out of Attorney Doe's world.

That is until three weeks later, when in a curt but polite letter written in the most formal of tones, John Smith dismissed Attorney Doe as both his personal attorney and company counsel. No reason was given other than the standard "philosophical differences" and the need for a change. Attorney Doe was dumbfounded and shocked, but after a time, he realized what had happened.

Through no fault of his own, Attorney Doe had become like the gummy candy, leaving behind a bad taste, which would always bring to Smith's mind the embarrassing and traumatic episode concerning his secretary.

While your experience with an attorney might not have involved issues quite so humiliating as that of John Smith's, there still may lurk in your unconscious a negative association with your lawyer. It might have been a criminal case, or a domestic dispute, or the filing for bankruptcy, or the airing of the family "linens" during the course of an estate battle over Grandfather's will. Many, if not most, matters requiring legal representation do not involve pleasantries.

So should you be reluctant to engage an attorney who otherwise appears to be the one for you and who performed admirably in the past, stop and evaluate the situation. If necessary, purge yourself of any undeserved aftertaste, and make your decision based upon what you have learned in this chapter.

On the other hand, you may discover just the opposite, that you cannot resist returning to the same lawyer time and again, despite a poor showing in "How To Score Your Lawyer" and a failure to meet the requirements of Chapter Three. This is a horse of another color, with ramifications which can be even more devastating.

Attorney Addiction:
Kick the Habit!

Do you find that you are unduly dependent upon your attorney? You may not realize it, but this is the case if any of the following applies:

You are no longer able to read and interpret the most basic of contracts, documents or letters.

You suffer from indecisiveness and are unable to make a decision without calling your lawyer. Such as, should you take an umbrella on a cloudy morning?

You are preoccupied with the legal ramifications of everything you do, even the most ordinary occurrence, such as whether to retain the ticket stub from a movie theater in the event you have to prove where you were that evening.

If more than three days pass without having spoken to your lawyer, you go into withdrawal, suffering abdominal pains, cold sweats, and delusions that you are being attacked by a band of cannibals bearing briefcases and brandishing giant gold plated ball point pens.

You enjoy paying your attorney's bills.

Should you display one or more of these symptoms or similar ones, then you are an Attorney Addict.

Unfortunately, there is no organized support group to wean you from this dependency. Yet the ramifications of this addiction can be deleterious to your health and your wealth, both of which diminish in direct proportion to the length of time this condition continues unabated.

If you don't believe this, consider the case of Maureen Masterson.

Maureen Masterson was one of those people you yearn to hate out of envy and jealousy, but have no choice but to admire and adore. She had both beauty and brains, having attended the Shipley School before graduating magna cum laude from the Wharton School at Penn and later earning her MBA at Harvard. With her family well-connected in financial circles, Maureen had little difficulty raising sufficient capital to run through a series of highly leveraged buyouts from which she amassed a small fortune, all of which was invested in a personal holding company for real estate and diversified enterprises across the country.

Maureen succeeded at any challenge she undertook. She could do no wrong and needed no help. So why consult an attorney when her marriage of thirteen years began to sour? Surely she and her husband Martin could work it out.

Negotiations started out well, but gradually deteriorated to the point where each became obsessed with who gets what, when, and how. During one particularly heated altercation, in an outburst of frustration, Maureen conceded more than she should have, and two days later Martin placed a settlement agreement under her nose.

Maureen had already ignored Chapter Two by not retaining an attorney when she needed one, and now with a time bomb ticking away, she rushed into hiring a lawyer without regard to Chapter Three. As a result, Maureen chose the wrong attorney, Sallie Slopgorf, a personal injury lawyer with the most garish ad in the Yellow Pages (Thou Shalt Not II), who supposedly knew all the right judges (Thou Shalt Not IV). To make matters worse, Maureen laid the foundation for a dependent relationship cemented with a sense of relief that she now had a lawyer who would take care of everything.

Preoccupied with the tumult in her life and worn from the emotional strain of the past several months, Maureen had neither the inclination nor the time to establish the ground rules in accordance with Chapter Four. This cost Maureen dearly. She got clobbered with the expenses (GR II) which included $6,000 for a very seamy looking private investigator who turned out to be Sallie Slopgorf's son-in-law; her elevated blood pressure was continuously aggravated by the many anxious hours and days spent waiting for Sallie's return calls (GR IV, responsiveness); and what began with at least some civility had now burst into a full scale blitzkrieg because Slopgorf went for the jugular (GR III, make your expectations known).

And if all this wasn't enough, despite her wealth, Maureen Masterson began to feel the pinch to the pocketbook with Slopgorf's monthly billings averaging $4,500. Not quite in keeping, she considered, with the "Oh, don't you worry about the fee my dear," the attorney nonchalantly smiled when they first met.

Maureen tried to get her lawyer to back off and be more reasonable, but it was too late. There was no way Maureen would ever be the boss of this operation (Chapter Five). She sat helplessly on the sidelines while Sallie took charge and set the course for the rest of Maureen's life. Finally, with an air of resignation, Maureen agreed to let Sallie do whatever she thought best, thereby relinquishing her last vestige of independence. Maureen Masterson had become a full-fledged Attorney Addict.

Attorney Addicts are unable to wind down and tie up the loose ends of their lawsuit (Chapter Six). So it went with Maureen Masterson. Her case dragged on for years—support hearings (sought by her husband), custody fights, disposition of marital assets (mostly hers), allocation of the substantial costs for the children's private schools and future college education, and finally, the actual divorce.

Whenever the subject of "liquidity" was brought up in the course of her financial ventures, Maureen always sardonically laughed to herself, thinking how her money seemed to be flowing from a wide-open faucet—funds to set aside for her children, dividing and then selling the assets, paying Sallie Slopgorf's outrageous bills, and on and on it poured. Through it all, Maureen behaved like an unwanted newborn, swaddled in diapers, grateful for the least bit of attention indifferently tossed her way by her surrogate mother, Sallie Slopgorf, Esquire.

Years of neglecting her businesses and investments, exacerbated by a downturn in real estate values and in the stock market, left Maureen cash-poor. She needed an attorney who could fend off creditors, renegotiate loans with the banks, and advise her on restructuring her assets and liabilities. Though at least this time she knew when to go to a lawyer, because of her dependency on her attorney, Maureen made the tragic mistake of returning to Sallie Slopgorf for representation. Attorney Addiction sealed the miserable fate awaiting Maureen Masterson.

As everyone but Maureen had expected, she lost everything. Yet up to the very end, with the filing of both corporate and personal bankruptcy, Sallie remained Maureen's attorney, garnishing another thirty thousand dollars in fees from the insolvency proceedings. All in all, during the four years Sallie represented Maureen, the insatiable attorney, much like a sponge, absorbed almost two hundred thousand dollars in fees plus whatever profit she skimmed from "costs incurred."

At the same time Slopgorf was redecorating her condo and spending winter vacations hobnobbing with the "beautiful people" in St. Barth's, Maureen lost her husband, all the marital assets she traded off to retain the holding company (destined to go down the tubes in bankruptcy), the affection of her children, who couldn't believe she had such a bitch for a lawyer, and even Ming and Ying her Siamese cats.

Slouched at a second hand roll-away style desk in her cramped study, contemplating putting an end to it all, Maureen's gaze fortuitously strayed to a tiny plaque mounted on the wall. It was inscribed with these words of wisdom written by Albert Camus, the French existentialist: "In the midst of winter, I finally learned that there was in me an invincible summer."

Freshly inspired and fortified with the belief she could still bounce back, Maureen Masterson sought out Sallie Slopgorf yet once more, this time in the hope her attorney could aid in finding a job.

Given her track record, ending in bankruptcy, Maureen's prospects in the financial world were slim to none for some time to come, Sallie pointed out. But, the attorney's thin lips stretched into a smile, Maureen was still a very attractive woman. That said, Sallie Slopgorf offered Maureen Masterson a position as a masseuse at a center city massage parlor in which Sallie was part owner.

Learn a lesson from Maureen Masterson and all the other Attorney Addicts. At the first sign you are growing overly dependent on your lawyer, kick the habit. Just say "No."

EPILOGUE

The Tortoise and The Hare, Part Two: The Untold Story

Everyone should be familiar with the children's fable about the tortoise and the hare, proving the moral that determination and deliberation will ultimately prevail over recklessness and overconfidence. Less known, however, is what occurred in the aftermath of the eventful race between the two competing critters.

For the hare to have graciously accepted defeat in the face of the embarrassment inflicted on him by the tortoise would have been totally out of character. Even more vindictive than a sore loser is a sore and humiliated one, which is what the bitter jackrabbit had become. The angry rabbit was obsessed with a burning desire for vengeance. He could think of little else and yearned for the day when he could make that tortoise stick its ugly head into its shell once and for all!

An array of revengeful fantasies raced through the rabbit's mind. Ultimately Mister Bunny hired an attorney to sue the turtle, for whatever reasons the crafty counselor could concoct. Money is no object, boasted the rancorous rabbit.

True to form, the poised and collected tortoise took no drastic action. Instead, our friend and hero of the tale pursued a path of cautiousness and wisdom. The first thing the turtle did was to buy and read this book. After that, it was a cinch, and you can finish the story for yourself.

The next time there arises even the remotest possibility you may need an attorney, re-read this book. Follow the example of the tortoise.

Lawyers are a lot like rabbits. Though sometimes appearing cute and cuddly, in a wink of an eye, they can be off and running with your money.

In parting let me say: Be firm and confident with your lawyer. You are the boss as long as you are the one paying the bills. Let your lawyer know this in no uncertain terms. Make it clear. "The buck stops here"—with you.